Thermocouple Temperature Measurement

Thermocouple Temperature Measurement

P. A. KINZIE
R. David Moore and Company

A Wiley-Interscience Publication

John Wiley & Sons

New York   London   Sydney   Toronto

Library of Congress Cataloging in Publication Data:

Kinzie, P. A., 1929 -
    Thermocouple temperature measurement.

    "A Wiley-Interscience publication."
    Bibliography: p. 239
    1. Thermocouples. 2. Thermometers and thermometry.
I. Title.

QC274.K5   536'.52   72-10121
ISBN 0-471-48080-0

Printed in the United States of America

10 9 8 7 6 5 4 3 2

Preface

The individual whose occupation involves science or technology usually acquires reference books for his field of interest. When he has a need for measuring temperature he may find that some of these books describe conventional techniques, including the method of thermoelectric thermometry. However, in the case of the latter subject only commonly used thermocouples are normally discussed, although there are many other couples which may be applicable for special uses. The purpose of this book is to present a summary of data and literature references for both well known and little used thermocouples; the intention here is to provide a convenient source of information for the study of unconventional requirements and their solutions.

There has been a proliferation of thermocouple types in recent years, and there have also been some suggestions for a greater degree of standardization in order to avoid chaos. I hope that this book will not inhibit the urge to standardize. The options of variety given here apply to specialized requirements that are few in number compared with the total of all thermocouple applications. On the other hand, this information will make it easier to compare alternatives, which will encourage the choice of a well known thermocouple in preference to the little known when such a choice is possible.

Thermocouples have often been categorized by the terms noble metal, base metal, and high temperature or refractory metal, and I have maintained this general approach to classification, as a glance at the table of contents will show. It is convenient, however, to include a separate chapter on mixed noble-base metal combinations and to include a section on nonmetal thermocouples. Performance characteristics are summarized for each type of thermocouple where such information is available. Data and discussion of the other physical properties of thermocouple materials are kept very brief and general. There is much more of the latter type of information in engineering and physics handbooks than can be included in a book of this size.

An introductory chapter is also included which briefly reviews some of the principles of thermoelectricity as applied to practical measurement problems. Also, the terminology used in the other chapters is defined and discussed in order to clarify concepts and terms as used here. A complete treatment of fundamental

v

principles is not attempted, because the reader will either have some general background in this subject or can consult an appropriate textbook or review article.

Finally, I have included a section in the Appendix which lists the more important thermocouples and their most distinctive characteristics, and which indicates those relatively standardized combinations that are in general use. Another section presents a list of review and general articles on the subject. All references are at the end of the book, and these can also be used as a bibliography.

A number of individuals and several organizations contributed assistance during the preparation of this book. Although I cannot list each contributor in a brief acknowledgement, I am most appreciative of their aid and consideration, and I welcome this opportunity to express my thanks.

Arcadia, California   *P. A. Kinzie*

Contents

Thermocouple Temperature Measurement

1

Outline of Fundamentals

SOME BACKGROUND AND PRINCIPLES

Thermoelectricity was discovered by T.J. Seebeck in 1821, at about the time that Ohm was conducting his studies of electrical resistance. According to Hunt (1964)[1] the first suggestion that the thermoelectric effect could be used for high temperature measurement was made by A.C. Becquerel within five years of Seebeck's discovery. Becquerel's measuring instrument was a magnetic needle held in the field of the closed loop carrying the thermoelectric current. He decided from his experiments that platinum and palladium were the most suitable combination for temperature measurement. Platinum is still widely used, both in measurement applications and as a reference material against which other materials can be compared. Palladium has been used to a limited extent.

A circuit of the type that Becquerel studied is shown in Figure 1-1a. A constant thermoelectric current exists in the loop, provided that T_1 and T_2 are held at constant unequal temperatures. The two conductors are assumed to be homogeneous but dissimilar materials.[2] The magnitude of the current may be large or small depending upon the electrical resistance of the circuit, but the Seebeck voltage which produces the current is determined by the temperatures of the two junctions where the materials interface, and by the materials' compositions. The direction of the current flow depends upon which material is thermoelectrically positive, and whether T_1 is warmer or cooler than T_2. The current is given an arbitrary direction in the figure.

It can be shown directly by experiment that the introduction of a third material C in the circuit, as in Figure 1-1b, does not change the thermoelectric voltage, provided that the new material is also homogeneous and that the new junctions are both at the same temperature. In principle there is no limitation on the temperature at other locations on C; the temperature can be varied without

1

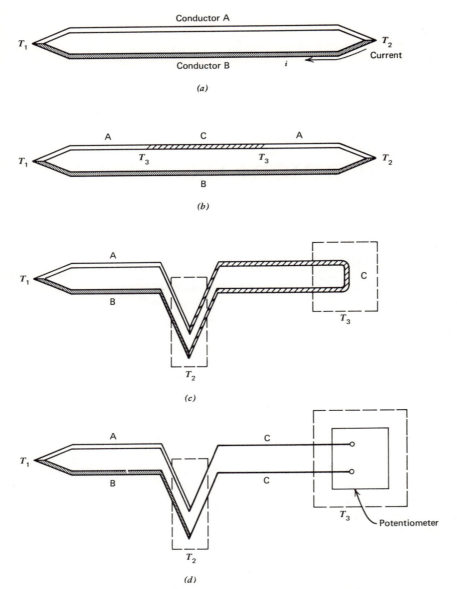

FIGURE 1-1 a. The basic thermoelectric circuit. b. Introduction of a third conducting material without changing the thermoelectric voltage. c. Placement of the third conductor between A and B without changing the thermoelectric voltage. d. A practical circuit for thermocouple voltage measurements.

disturbing the thermoelectric voltage. The current in the circuit is changed, of course, by any alteration of the circuit resistance. The circuit in Figure 1-1b can be generalized for any number of additional materials, and it can also be shown that the location of C is unimportant relative to the two original junctions. Figure 1-1c is an example of the third material introduced between A and B with junctions at temperature T_2, while a portion of C is held at any arbitrary temperature T_3.

These results are really consequences of three basic demonstrable relations, which are usually called the three laws of thermoelectricity. Discussions of the three laws have been given in work by Roeser (1941) and by Finch (1962); the latter reference describes the evolution of the circuit with three dissimilar conductors, using fundamental principles. A similar discussion with applications to practical thermocouple circuitry is given in the manual on thermocouples by the American Society for Testing and Materials (1970, p. 12).

A practical circuit is shown in Figure 1-1d for measuring the thermoelectric voltage generated by A and B, with the junction temperatures at T_1 and T_2. The temperature T_2, called the reference temperature, must be known if T_1 is to be determined, since the thermocouple voltage is a function of the difference between T_1 and T_2. A simple yet satisfactory method of obtaining a constant known T_2 is to maintain the reference junction in a bath of melting crushed ice and water, well stirred and contained in a Dewar flask.[3]

The circuit in Figure 1-1d can be used for comparing the thermoelectric output of material A with platinum as B. By maintaining T_2 at the temperature of 0°C, and exposing the other junction at known temperatures T_1, the user can construct a graph or table of thermocouple voltage versus temperature. New materials are often evaluated against platinum in this manner, although other reference metals are sometimes used.

The difference between the thermocouple voltages of two materials at a given hot junction temperature, each relative to platinum, gives the voltage of one material relative to the other. To demonstrate this, Figure 1-2 shows conductors A and B referenced to platinum so that the separate voltages V_a and V_b relative to platinum can be measured. The measuring instrument can next be connected to terminals A and B, obtaining the voltage across the circuit consisting of conductor A, the platinum conductor, and conductor B. Now the platinum is present only as a third material, both of whose junctions are at the same temperature, so the voltage at terminals A and B must be that of a thermocouple constructed of A and B alone. The output is $V_a - V_b$. If $V_a = V$ and $V_b = -2V$, both relative to platinum, the net output is $V - (-2V) = 3V$.

THERMOELECTRIC POWER

Pairs of materials are chosen for thermocouples based on a number of criteria for usefulness. One such requirement is that of a usable thermoelectric voltage-temperature output characteristic over the desired range of temperature measurement. The sensitivity or thermoelectric power is a measure of whether a pair of materials is useful or not in this respect. The thermoelectric power S of a couple is the rate of change of output voltage with respect to temperature at the measurement junction, in short, the derivative of voltage with temperature. It can be found by taking the slope of a voltage-temperature curve, or by directly measuring the small voltage change δV generated by the circuit in Figure 1-1d, when T_1 is increased to $T_1 + \delta T$ and T_2 is held constant. The value of S at a given temperature is then approximately

$$S \approx \frac{\delta V}{\delta T}$$

or, in the limit of very small δT, by the exact relation

$$S = \frac{dV}{dT}$$

Given that the temperature T_2 in Figure 1-1d is a fixed reference temperature, the sensitivity of a couple is the thermoelectric power at the junction with temperature T_1. Thermocouples with values of S once considered unacceptably low for most applications can now be used with satisfactory results, because of the high sensitivity of present-day measurement instruments. Couples with thermoelectric powers of only a few microvolts per Celsius degree are not uncommon.

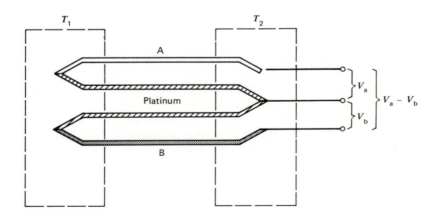

FIGURE 1-2   Relationships between thermocouple voltages relative to platinum and to each other.

The magnitude of S almost always changes with temperature; in other words, sensitivity depends on temperature. There are few linear thermocouples because of this fact, unless the range of measured temperatures is deliberately made very small. The voltage-temperature curves of some pairs of materials display maxima and minima and may go through zero voltage with consequent change in polarity. Normally such extreme behavior is sufficient to eliminate those couples from practical use. Exceptions have occurred where other properties such as resistance to chemical attack are outstanding.

Representative values of thermoelectric power are given in the discussion for most thermocouples covered in Chapters 2 through 5. In many instances no information was available other than voltage-temperature curves, so numerical values were obtained in such cases by graphically determining the slope of the curve at a given temperature, or by taking an average slope over a range of temperatures. The resulting data, generally expressed to one or two significant figures, are of low accuracy. Nevertheless, the results are included on the grounds that they are helpful in estimating and comparing thermocouple sensitivities for application work.

OTHER PROPERTIES OF PRACTICAL THERMOCOUPLES

The thermocouple circuits in Figure 1-1 were discussed in terms of ideal materials, each of which had identical properties at all points. Yet, even the best real materials are not everywhere identical, and wherever a significant deviation from the average occurs the result is the equivalent of inserting a section of dissimilar material into the circuit. Therefore a real material may generate an error voltage along its length whenever a temperature gradient is present. Fenton (1971) discussed the implications of this type of condition and showed that the concepts of the laws of thermoelectricity in terms of ideal materials can be misleading. Fortunately, the errors introduced by nonideal materials are often small enough to make practical measurements possible.

Differences in thermoelectric properties are caused by both chemical and physical factors. Chemical differences exist when the as-prepared composition varies from point to point, or where regional contamination or component depletion occurs later. Physical differences result when the crystalline structure of the material varies, which can be caused by the initial wire drawing and heat treatment, work hardening during handling, or unequal amounts of high temperature exposure during use.

Powell, Caywood, and Bunch (1962), Hust, Powell, and Sparks (1971), and Sparks and Powell (1971, p. 19) have discussed these effects in terms of different categories based on the length of conductor involved. A somewhat similar system is used here where short range effects over distances up to some

20 cm are called inhomogeneities, and medium or long range effects are both classified as nonuniformities.

Inhomogeneity

In Figure 1-1b, let the length of material C represent an inhomogeneity in conductor A. The error contribution due to C is zero when each end is at the same temperature T_3; however, when a difference in temperature ΔT exists between the two ends an error voltage is developed which is proportional to ΔT.[4] The magnitude of the error also depends on how much the thermoelectric power of C differs from that of conductor A. This is a simple concept, yet it is surprisingly difficult to obtain meaningful practical information on materials and to apply the information during their use.

Homogeneity studies have been made of the more common thermocouple materials, but most information results from simple comparative experiments, from which the most homogeneous thermocouple is selected from several types for a given set of conditions. Since there is little detailed information on most materials, only comparative type inhomogeneity data will be included here; the term inhomogeneity will be applied to effects in conductor lengths of some 20 centimeters or less.

Inhomogeneities may be present in newly received wire, but very often they appear or become more noticeable in use owing to various contamination processes or simply because of the unavoidable difference in the amount of heat treatment that sections of the conductors receive during use, depending on their location.

The inhomogeneities that develop during use may affect calibrations made after removal from service. Unless the temperature profile along the wires closely matches that of prior service conditions, a post-service calibration after removal may not give the same results as an in-situ calibration. For example, if recalibration is conducted with the inhomogeneous region produced by nonuniform heating now located in a low temperature gradient region, errors that had developed during use would no longer appear during recalibration. As a result, the thermocouple would be mistakenly judged as satisfactory. This has led to the general recommendation that, once installed, couples should not be moved, and where possible should be calibrated in place. Otherwise, calibrations are best made before installation.[5]

Uniformity

Medium and long range variations in materials properties generally affect the thermoelectric characteristics of couples made from a long length of wire, or of couples made from different lots or batches of materials. In the case of non-metal couples there appears to be a similar variation between rods or tubes from a given batch of raw materials and between conductors from different batches. Thin film couples made during the same run in a deposition process can be compared with each other or with similar results from a different run. As might be expected, whatever the fabrication process, couples from the same batch are generally in better agreement than those from different batches. All of these characteristics will be discussed in terms of uniformity, where the latter term refers to the per cent error in output at a given temperature, compared with that of a standard or with an average of several samples.

Instability and Drift

The output voltage and the thermoelectric power of a thermocouple may change over a period of time, and such changes are caused by one or more contributing factors. A very common occurrence is a change in the degree or the nature of inhomogeneities in a region of steep temperature gradient. Regardless of cause, short period changes in the output of a thermocouple are called instabilities. The term will be used here for changes occurring in a period of a few minutes to an hour. Other terminology in general use includes such words as short term drift, shift, and fluctuation. Drift is defined here as long period output change that occurs over exposure periods of more than an hour; this has also been called calibration shift, decalibration, or simply as a change in calibration.

Repeatability

This term, and precision, or reproducibility, are often used interchangeably to describe the error observed when a given calibration point is repeatedly measured, or when a previous set of conditions is reestablished to produce a temperature that may not be an accurately known quantity, but which is a closely repeatable condition. Repeatability is defined here in terms of the difference between a single measurement and the average value of a number of similar measurements of the thermocouple voltage for the given temperature condition. The result is expressed as a percentage of the average voltage or as an apparent temperature difference. In some practical applications where recalibration is not feasible, a measurement of repeatability is the only criterion of whether a thermocouple is performing satisfactorily or has deteriorated.

It is evident that instability will lead to decreased repeatability, and if a considerable period of time elapses between measurements, drift also affects the results. Therefore the three quantities are often interrelated. However, if thermocouple stability is high and measurements are made over a short time period, repeatability can be used as a measure of the performance of the thermocouple or of the entire measuring system, and of the ability to precisely repeat a calibration point or an arbitrary temperature condition.

Accuracy

Thermocouple accuracy in the general meaning of the term is the total error in measuring a given temperature, expressed as a per cent of that temperature (or as a per cent of the maximum rated thermocouple temperature, calibration temperature, or application temperature, depending upon how the term is used in a discussion). Accuracy may also be used in referring to the absolute error in Celsius degrees or in kelvins, in measuring a given temperature. At best, thermocouple accuracy is such that temperatures can be measured to within ±0.1 or ±0.2°C with certain types of thermocouples, following a precise laboratory calibration, and with optimum application conditions. Unfortunately, this degree of accuracy is often expensive in terms of time and required equipment; on the other hand, such accuracy is seldom required for application work, so that more economical but adequate procedures can be used.

It is evident that a user can choose to relax the requirements of the calibration process and use less accurate but more convenient calibration equipment, take fewer points, and use simpler curve fitting procedures. Even more important is the option of calibrating only a few of the total number of thermocouples used. This requires that the user establish, or have available, information on the uniformity of a given batch of wire and of batch-to-batch uniformity if necessary.

The need for convenience and economy for thermocouple use in industry has led to the establishment of standard calibration reference tables for the most widely used combinations. The tables are based on an average of calibration data from a number of very precise measurements, often using samples from several different suppliers. Therefore, a table is not a calibration result for a single thermocouple but is a representative set of data which can be matched within a certain error band by the user's thermocouples. The suppliers of commercial thermocouple materials use manufacturing and quality control processes to insure that materials and thermocouple probes will continue to match the reference tables to within a specified accuracy, once the tables have been published. Typical accuracies for moderate temperatures are ±3/4 per cent with respect to the indicated temperature, for a standard tolerance rating, and ±3/8 per cent for what is called reference, premium, or special grade material. This same commercial wire can, of course, be calibrated on a couple-by-couple basis with consequent improvement in accuracy.

A method of obtaining a moderate degree of accuracy without a complete calibration program is to construct a difference curve with respect to the reference table. A relatively small number of accurate calibration points determined by the user establishes a curve which is then compared with a curve constructed from the reference table. The difference between these curves serves as a first order correction of data from the calibrated thermocouple.

Response

Transducers are usually rated in terms of their response to a specified transient input of the quantity which is to be transduced (converted to an electrical or otherwise measurable output). Thermocouples are sometimes rated in terms of how rapidly the thermocouple temperature rises to some fraction of the temperature of the medium to which they are suddenly exposed. Ideally, the result would be the time response to a step input of temperature, but because of heat transfer conditions it is apparent that the response to immersion in boiling water is faster than response to still air at the same temperature. Although this type of test often gives the user an idea of what to expect, the factor which is important in many applications is not the time response of the thermocouple alone but the change in the time response of the specimen caused by its presence. If this change is small, and if the thermocouple temperature is not unacceptably different from that of the specimen after some initial period of time, the user obtains adequate system response. It is true nevertheless, that a fast immersion response time is a favorable indicator of performance possibilities.

Response will not appear here in the discussion of thermocouple properties unless it is a major influence in choosing a particular type of couple. There are a few cases where a thermocouple may be suited for a special application because its thermal properties permit an optimized design for good response.

COMMONLY USED THERMOCOUPLES

Much of the temperature measurement work with thermocouples can be adequately done by one of a few standard types. Consideration of one of these types is often advisable, even when the expected performance may not be quite as satisfactory as with a custom selection. The reason for this, of course, is that extra expense and labor may be avoided. The commonly used materials are readily available as thermocouple stock of known quality and characteristics. Also, the backlog of application knowledge is much greater than for little used combinations. A summary of thermocouple types and applications is given in the Appendix, and those couples which are most often used in the United States are specifically designated.

THERMOCOUPLE REQUIREMENTS FOR ENVIRONMENTAL EXPOSURE

The performance of a thermocouple deteriorates under many application conditions. In fact, no one thermocouple has been found which performs successfully for all common usage requirements. Instead, a compromise must be made in selecting the most satisfactory of a set of candidates for a particular requirement. The limitations of thermocouples make it desirable to categorize and describe the principal environmental considerations in terms of temperature range, atmospheric protection, contacting materials, and mechanical limitations. Each of these factors is further discussed below.

Temperature Range

The performance of thermocouples varies greatly with temperature. Many materials that are usable at high temperatures because of their high melting points are less satisfactory than moderate melting point substitutes at lower temperatures. The requirements for performing in a certain temperature range, together with some of the other environmental factors, have led to a de facto classification of thermocouples for cryogenic use, general use (moderate temperatures), or high temperature use. This is a convenient classification because quite often a user is interested in only one of the three ranges and can usually select a pair of materials that is adequate. If operation for both cryogenic and moderate temperatures or moderate and high temperatures is required, some definite compromises in performance must be accepted.

Atmospheric Protection

The effects of the atmosphere are usually important at moderate-to-high temperatures, since chemical reactions and diffusion processes take place rapidly under such conditions. An inert atmosphere is necessary in many cases, especially for long term operation; vacuum can be used as an alternative, but vaporization of the thermocouple materials may become significant at higher temperatures. Some of the noble metals have excellent characteristics in oxidizing atmospheres at moderate-to-high temperatures, but neither these nor many of the base metals are generally satisfactory for reducing atmospheres. These restrictions make it necessary to isolate the thermocouple from the environment for many types of service. Thermocouple wells, probes, or sheathed cables can be used. In the latter cases the hot junction can be either grounded to the inner surface of the case or sheath, or left ungrounded but supported by insulation for more complete isolation.

Contacting Materials

Thermocouples usually require mechanical support and electrical insulation. Contact with an additional material is often a prerequisite for obtaining a true temperature measurement of the material itself rather than of the general surroundings. All of these materials must be compatible with the thermocouple to avoid error by progressive contamination. Such contamination is sometimes a major problem at high temperatures, where chemical reactions may begin or become significant in terms of reaction rate. Furthermore, metallurgical processes such as interdiffusion between metals may occur. Also, the electrical resistivity of the insulators degrades. These considerations must be kept in mind when designing or applying thermocouples at high temperatures, particularly at and above 1600 to 1800°C.

At lower temperatures where organic materials or plastics can be used as insulators, there are few limitations. However, in the case of bare wire couples, chemical reactions with corrosive media must still be considered in some applications.

Mechanical Limitations

One of the most striking advantages of the thermocouple as a temperature sensor is the small size of the junction and wiring that can be attained. Its adaptability for installation in confined areas, small assemblies, and remote or virtually inaccessible locations is almost limitless. There is variety available in the conductor and junction designs themselves; conductors can be cables, wires at least as small as 0.02 millimeter, or as thin films on a substrate. In regard to physical size and adaptability for special requirements, there are probably fewer limitations with thermocouples than with any other form of contact-type temperature sensor.

In harsh mechanical environments where vibration, shock, thermal expansion, or structural loading are important factors, the thermocouple installation must be appropriately designed. In general, circumventing this type of limitation is similar to solving typical structural design problems. There are a few situations, nevertheless, where design does become very specialized. For example, the selection of materials and the fabrication process for small-diameter metal sheathed thermocouple cables was a challenging problem that manufacturers have successfully met. Still, the user may have less than optimum performance if he requires a certain sheath material, and attention is not given to such possibilities as dissimilar thermal expansion coefficients of the sheath, the insulation, and the wires.

One electrical characteristic of an installed thermocouple may be very dependent upon size limitations and other mechanical design features. This is the response time to a sudden temperature change. The final size chosen for a thermocouple is sometimes a compromise between a very low mass installation to obtain rapid response and a more rugged installation to improve vibration and shock tolerance.

EXTENSION WIRES AND ACCESSORIES

Thermocouple wires, either bare or insulated, are expensive when ordered in large quantities. One method of economizing, particularly with noble metal thermocouples, is to use extension wires. These wires are selected pairs of alloys which have a net thermocouple voltage very similar to that of the thermocouple itself over a limited range. Such wires cannot be used in the high temperature portion of the circuit. They can be spliced into the circuit between the moderate temperature region and the reference junctions. The extension wire alloys are base metals which are less costly than the high temperature thermocouple materials. This becomes particularly important when long lengths of conductor are required to reach the reference junctions.

The extension wire alloys are chosen to match the characteristics of the thermocouple between about 0°C and an intermediate temperature of not over a few hundred degrees Celsius at most. The upper limit depends upon the reactivity of the extensions with the atmosphere and on the type of insulation selected, as well as the temperature at which the extension characteristics begin to deviate significantly from the thermocouple characteristics. Extensions are available for some of the base metal couples as well as for the noble metals. Such extensions for base metal pairs are often lengths of wire of the same nominal composition as the thermocouple, but which do not meet the required specifications for high temperature output. Helpful application information on extension wires has been given in the work by Starr and Wang (1969, 1971).

Connectors, terminals, and other hardware are available that are made of commonly used thermocouple materials or of comparable alloys. The combination of the thermocouple wires, extensions, and appropriate hardware permits a nearly homogeneous circuit from the hot junction to the reference junction. Copper conductors are usually used on the measuring side of the reference junction, and solder alloys can be purchased that have thermoelectric powers comparable with that of copper, for minimizing locally generated thermoelectric voltages.

A survey of thermocouple hardware is included in the manual on the use of thermocouples by the American Society for Testing and Materials (1970, p. 55). Discussions of thermocouple junctions, terminations, insulators, connectors, and protection tubes are included, with an extensive table on recommended protection tube materials for specific applications. Much information on sheathed ceramic insulated thermocouples is also presented. Another publication, particularly for industrial and general purpose requirements, is ASA C96.1-1964 by the American Standards Association (1964).

USE OF THE REFERENCE SECTIONS

Thermocouples are classified by composition in the following chapters, under noble metals, mixed noble-base metals, base metals for low and moderate temperatures, and, finally, base metal and nonmetal couples for moderate and high temperatures. This arrangement is similar to the classification sometimes found in thermocouple review articles. There are also subcategories in each chapter because of the large number of entries which are listed. This is a hybrid classification scheme involving both composition and temperature range. The arrangement is not an artificial one, since the properties are interrelated.

Information on a thermocouple can be located by examining its nominal composition to determine which category it is in and therefore which chapter is applicable. The list of elements in Appendix B can also be used as an alternate basis for searching. For example, under rhodium, the page numbers referring to all thermocouples containing that element are given.

The nomenclature used for describing thermocouple nominal composition requires some explanation. An alloy is specified by the chemical symbols of its constituents. The weight percentages of the solutes are given by symbol prefixes. For example, platinum with 10 weight per cent rhodium is written at Pt-10Rh. Atomic weight percentages are used in a few cases and the abbreviation at. is used to denote that fact; gold with 0.07 atomic per cent iron is written as Au-0.07at.Fe.

A single sign convention for thermocouple polarity is followed throughout to avoid confusion. Most moderate and high temperature usage results in polarity being determined by a measurement at the reference junction wires and with the reference junction cooler than the measuring junction. Here the positive conductor is listed first so that the nomenclature Pt-10Rh vs Pt signifies that Pt-10Rh has positive polarity at the reference junction.

Some cryogenic specialists will not feel at home with this notation because in their applications the reference junction may be warmer than the measuring junction, leading to a reversal of the observed polarity. Nevertheless, in order to be consistent and avoid confusion, the convention of determining polarity by voltage measurement at the cooler junction will be maintained for all couples, regardless of whether they are normally used at high temperatures or at cryogenic temperatures.

Thermoelectric voltages are customarily given for a reference junction at $0°C$ unless the reference junction temperature is otherwise specified. That custom is followed here. Cryogenic thermocouple data are sometimes given with a 0 K reference, or with some other cryogenic temperature reference. Information of this type is included where applicable.

The thermocouples listed are those which have been used for measurements or that were investigated and reported in measurements-oriented papers. There is also much information in the literature of physics on thermoelectricity, but the physics-oriented paper does not usually include the type of applications information stressed here. This is also true of the studies of thermoelectricity for electric power generation and for refrigeration. Most of the resulting thermoelectric data have been excluded here, except where the information can clearly apply to measurement applications.

The voltage-temperature curves which are given here vary widely in accuracy depending on the original source. These curves are quite adequate for comparing general characteristics, but they should not be used in lieu of calibration tables or larger scale calibration curves. Also, as was mentioned earlier, numerical values of thermoelectric power were determined graphically in many instances and are representative, but of low accuracy; these are presented in terms of one or two significant figures.

FOOTNOTES

1. All references, in alphabetical order, will be found at the end of the book.
2. Real materials are never perfectly homogeneous; hence this assumption applies, strictly speaking, to an idealized condition.
3. The two thermocouple junctions are often called the hot and cold junctions, but the cold junction is not always the reference junction, even though common usage sometimes suggests that relationship.
4. This is true provided that C is itself homogeneous. Short segments of real materials may approach this condition. Summation techniques can be used to simulate longer lengths.
5. Even small changes of wiring placement during use may result in changes of output. Studies of this type of error have been made in terms of the change of immersion depth of a thermocouple in a special furnace or other heat source, and of the resulting effect on the thermocouple voltage.

2

Thermocouples of Noble Metal Composition

PROPERTIES OF THE NOBLE METALS

The family of elements consisting of silver, gold, and the six platinum metals (ruthenium, rhodium, palladium, osmium, iridium, and platinum) is often called the noble metals group. The classification is not a rigorous one and varies somewhat in the literature of chemistry and metallurgy. All of these elements are relatively inert, compared with metals in general, and the platinum metals resemble one another in many other respects.

Silver, gold, and platinum are well known as soft, ductile metals, with high electrical conductivities. Palladium, which is less well known, is also ductile, but rhodium, iridium, ruthenium, and osmium become progressively harder in that order. Rhodium and iridium can be worked, and wires of these metals are sometimes used in thermocouples. Ruthenium is very difficult to work and the problem of preparing even short lengths of wire or rod has greatly hampered the investigation of its electrical properties. Osmium has been prepared in the form of bars by powder metallurgical methods, but not in the form of drawn wire. All of these elements are expensive, because of their rarity.

The softness of silver, gold, and platinum is a disadvantage when structural strength is required. Hard drawing would improve their strength to some degree, but recrystallization and softening occur at temperatures well below the maximum application temperature. These and other metals are usually calibrated and used in a heat-treated condition to eliminate possible changes in thermoelectric characteristics which might otherwise occur. A commercial process of drawing a composite assembly of fine platinum wires into a single wire, called Fibro platinum, has been described by Hill (1961, 1962)[1], and this improves the strength considerably. Successful work of similar nature with other soft metals has not yet been reported.

Representative data on thermocouple voltages referenced to platinum are shown in Table 2-1. The elements are listed in the order of increasing atomic weight. It can be seen from the table that palladium is the only noble element that is thermoelectrically negative to platinum. Silver and gold, on the other hand, are the most positive. Gold has a slight edge over silver below about 400°C. Above this temperature, silver becomes increasingly positive relative to gold.

TABLE 2-1
Thermoelectric Data for the Noble Elements Referenced to Platinum

Element	Voltage (mV) at Temperatures Shown					
	200°C	600°C	1000°C	1200°C	1400°C	1500°C
Ruthenium[a]	- - -	3.994	9.760	13.589	- - -	- - -
Rhodium[b]	1.61	6.77	14.05	18.42	23.00	25.35
Palladium[b]	−1.23	−5.03	−11.63	−15.89	−20.41	−22.74
Silver[b]	1.77	8.41	(16.20 at 900°C)			
Osmium	(Not available. Thermoelectric power[c] of +12 μV/°C at 1000°C)					
Iridium[b]	1.49	6.10	12.59	16.47	20.48	22.50
Gold[b]	1.84	8.12	17.09	- - -	- - -	- - -

[a] Anon. (1965).
[b] Roeser and Wensel (1941, p. 1293).
[c] Estimated from curves of absolute thermoelectric power by Vedernikov (1969), and converted to a platinum reference.

The addition of another noble metal to platinum, forming a binary alloy, usually improves the mechanical properties of the metal as the alloying element percentage is increased from a nominal value up to a few per cent. In some cases, such as with rhodium, useful alloys result over virtually the entire composition range. On the other hand, amounts of ruthenium or osmium of more than a few per cent result in brittle platinum based alloys. Gold and silver have been little used in binary platinum alloys for thermocouple work, because of the resulting decrease in the alloy melting point without significant improvement of other properties. Thermocouple use of unalloyed gold and silver is limited because of the low melting points and the high costs of these elements, compared with base metals which can often be satisfactorily substituted. Silver is sometimes used instead of copper for improved performance in oxidizing atmospheres.

The platinum metals show some reactivity in typical use conditions, regardless of their reputation for inertness. Even platinum oxidizes to a minor degree on the surface when exposed to air in the 600°C temperature region. Rhodium shows more pronounced tendencies to surface oxidize. At high temperatures there is sufficient surface oxidation of iridium to produce a visible black "smoke."

Ruthenium oxidizes so rapidly that for thermoelectric applications platinum-ruthenium alloys must be used in an inert atmosphere above about 1000°C. The same is true of osmium alloys above a few hundred degrees Celsius.

Platinum metals volatilize in a vacuum, and there has been some controversy about the oxidation-volatization processes which take place at the surface of platinum-rhodium alloys in various atmospheres. Nevertheless, most noble metal thermocouples can be used in air, atmospheres of helium and the other inert gases of chemistry, or in nitrogen. Use in vacuum has been found to be satisfactory for short periods, and, where partially confined within insulators, for long periods of time.

Pure carbon dioxide does not affect the platinum metals. The same is true for carbon monoxide with respect to platinum, rhodium, and iridium. Palladium, however, becomes embrittled and ruthenium may also react, at least under some temperature-pressure conditions. No embrittlement of platinum metals by pure hydrogen has been reported, although palladium absorbs large quantities of the gas and can transmit it by diffusion.

Unfortunately there are a number of typical application problems where even trace amounts of environmental contaminants or noble metal impurities result in reactivity of the platinum metals in hydrogen and other reducing atmospheres. In spite of the demonstrated inertness of these elements in such atmospheres in the laboratory, noble metal thermocouples can be used in reducing atmospheres only under special conditions. Much of the older literature on this subject strongly recommended against such usage. Other gases also affect these metals. The reactivity with gases of the halogen group is well known, and at least some of the platinum metals are affected by sulphur dioxide.

In terms of reactivity with liquids, the platinum metals are inert in a wide range of chemicals commonly encountered in engineering. Most of the elements are even inert in the strongest single acids. (Palladium is attacked by nitric acid.) Platinum and palladium are more readily attacked by acid mixtures such as aqua regia than are the other members of the platinum family. Generally speaking, little measurements-oriented work has been done on liquid exposure; chemistry texts are a good source of information on properties of the platinum metals in liquids. On the other hand, the effects of solids are of considerable importance for measurement application. Metallurgical considerations are often dominant in this case.

Platinum and its alloys are usually inert with respect to the more refractory oxides, although some exceptional conditions were described by Darling and coworkers (1970, 1971) in a series of four journal articles and a conference paper. It was shown that typical good thermocouple performance resulted for oxygen levels as low as 50 ppm in an argon atmosphere, but reactions with metal oxides occurred when a gettered argon atmosphere or a vacuum with much smaller amounts of oxygen was used.

Carbon and the platinum metals are nonreactive when both are very pure. Where reaction has been reported, it has very likely been due to presence of impurities or contaminants. Contact with easily reducible metal compounds should be avoided because of possible alloying of the reduced metals with the noble metals. Direct contact of platinum metals with base metals can result in error due to contamination by diffusion of base metal atoms into the noble metal. This has been found to be true of nickel, for example, as well as for low melting point metals such as lead or bismuth. When gross contamination occurs, the alloying of the two metals can result in a decrease of the melting point and changes in the metal properties. Contact with base metals is sometimes required to establish a meaningful relation between the thermocouple junction temperature and the base metal temperature, but the compromise that must be accepted in terms of progressive error should be recognized.

The catalytic properties of the platinum metals are usually an undesired and error-producing feature, as far as thermocouple temperature measurements are concerned. Formerly it was thought that temperature rises caused by catalytic reactions in the local region near a thermocouple were not significant. This is not the case for some conditions, however, particularly when the temperature is raised to the region near the ignition point of the reactants. Olsen (1962), Stanforth (1962), Alverman and Stottman (1964), Stern, et al. (1970), and Thomas and Freeze (1971) have all contributed information showing that several of the platinum metals and their alloys can cause serious errors by means of catalysis. These errors cannot be easily predicted because a number of paramaters affect the results.

SPECIAL PROPERTIES AND TREATMENT OF NOBLE THERMOCOUPLE METALS

It is important that these metals be uniformly pure, although it may not be necessary or even desirable to require purity of the highest available grade. Table 2-2, using data from Zysk (1964), gives representative figures for some of the noble metals.

The pureness of platinum was judged for many years by its thermoelectric behavior relative to a length of pure platinum wire at the National Bureau of Standards, called Platinum 27. Wichers (1962) described the circumstances behind the adaptation of that particular sample of wire for a standard. More recently, an improved platinum standard, known as Platinum 67, was adopted at NBS. This was discussed by Burns (1971).

With the one exception of gold, an element present as an impurity in platinum increases the thermoelectric power of a thermocouple consisting of the impure metal and Pt-67. According to Rhys and Taimsalu (1969), the increase is linear with respect to the weight per cent of impurity up to at least

1000 ppm or 0.1 per cent. Therefore, a relatively high thermoelectric power is indicative of low purity.

TABLE 2-2
Representative Purity Data for Thermocouple Grade Noble Metals[a]

Platinum	Standard Grade	99.99 per cent (minimum)
		$\alpha = 0.003910$ (minimum)[b]
	Reference Grade	99.995 per cent (minimum)
		$\alpha = 0.003923$ (minimum)[b]
Rhodium	Standard Grade	99.95 + per cent
	Reference Grade	99.99 + per cent
Palladium		99.99 + per cent
Gold	Cryogenic	99.999 + per cent
	High Temperature	99.99 + per cent
Silver	Cryogenic	99.999 + per cent
	High Temperature	- - - - - - -
Iridium		99.95 per cent (minimum)

[a]Information selected from Zysk (1964).
[b]This quantity is discussed in the text.

Caldwell (1962) reviewed a different method of purity determination. This criterion of purity is the magnitude of the ratio

$$\alpha = \frac{R_{100} - R_0}{100 R_0}$$

where R_{100} and R_0 are the electrical resistances of a sample of wire at 100°C and 0°C, respectively. The value of α increases with increased purity, and it appears that the ultimate value corresponding to zero impurities is near 0.003928. Values of α for standard and reference grades of platinum are given in Table 2-2. The quantity

$$\frac{R_{100}}{R_0} = 100\alpha + 1$$

is also used in this respect, and its maximum value is 1.3928.

Spectrographic analysis can also be used to determine purity, and this was discussed by Rhys and Taimsalu, who indicated that the method is satisfactory for preliminary selection of thermocouple grade platinum. It was shown that when all impurities are below the spectrographic limit for estimating their presence in parts per million, then the value of R_{100}/R_0 is at least 1.39235. This corresponds to an α of 0.0039235, a figure slightly above the minimum value given for reference grade platinum in Table 2-2, and well above the minimum for standard grade platinum. This, of course, indicates a good degree of purity.

Effects of impurities on the value of α were studied by W. F. Roeser, as later reported by Corruccini (1951), who also discussed the subject. Rhys and Taimsalu (1969) verified earlier evidence that some impurity elements produce much greater thermoelectric changes than others, when compared on the basis of either an equal weight per cent or equal atomic weight per cent of solute added to platinum. Cochrane (1970, 1971) described work which resulted in a generalized relationship for thermoelectric voltage due to a platinum impurity in terms of its atomic weight and several other parameters.

Work hardening, recrystallization, and strain also affect thermoelectric power, and so it is good practice to heat treat a thermocouple up to a temperature above the maximum use temperature, if possible, prior to calibration. Corruccini (1951) discussed annealing procedures for platinum, and Zysk (1962) reviewed work on the subject. McLaren and Murdock (1971) showed that relatively lengthy treatment is required for achieving the best possible results. Heat treatment has also been used with other noble metals and alloys, although in many cases the metal embrittles rather than softens.

ELEMENTAL AND BINARY PLATINUM ALLOY THERMOCOUPLES

Gold and Platinum

Au vs Pt.   Sirota and Mal'tsev (1959) investigated thermocouples of gold and platinum and compared them with platinum-rhodium couples[2] during tests in the 190 to 550°C temperature range. Repeatability was 0.2 μV (0.013°C) or better for Au vs Pt compared with 1 μV or 0.1°C for platinum-rhodium.[2] Homogeneity was also determined and it was found that temperature errors caused by inhomogeneities were five times smaller for Au vs Pt than for the platinum-rhodium type thermocouples. Data from Sirota and Mal'tsev are included in the curve in Figure 2-1.

Merryman and Kempter (1965) chose this combination for the measurement of temperature in an apparatus to determine the thermal expansion of polycrystalline materials by X-ray diffraction techniques. The calibration range was 0 to 1000°C and the couple was used in a vacuum. Calibration data are shown in Figure 2-1. The thermoelectric power is 17.5 μV/°C at 500°C and 25.9 μV/°C at 1000°C.

The combination of gold with platinum is a good example of the concept that pure elements generally provide the highest uniformity of thermoelectric characteristics. Here there is no dependence on maintaining a specified alloy composition or, alternatively, depending upon a selection process where wires have been pre-chosen such that the voltage characteristic remains within certain deviation limits. This uniformity can be seen from Table 2-3, which compares a calibration from the compilation of thermoelectric properties by Roeser and Wensel (1941, p. 1293), and the independent results of the two references discussed here.

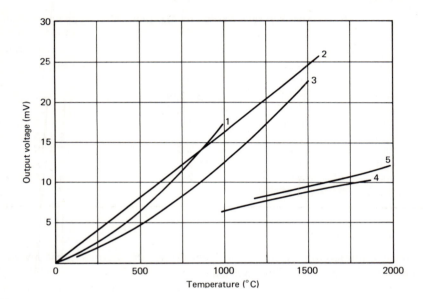

FIGURE 2-1   Platinum thermocouples with gold, iridium, and platinum-iridium alloys. 1. Au vs Pt, Sirota and Mal'tsev (1959); Merryman and Kempter (1965).   2. Pt-10Ir vs Pt, Bennett (1961). 3. Ir vs Pt, Roeser and Wensel (1941, p. 1293).   4. Ir vs Pt-20Ir, Troy and Steven (1950).   5. Ir vs Pt-30Ir, Troy and Steven (1950).

Except for the figures at $1000°C$, a temperature rather close to the melting point of gold at $1064°C$, the agreement in the table is within 13 μV. On the other hand, the relatively low melting point of gold is a disadvantage and limits the range of this combination. The high thermal conductivity of gold, plus a tendency to acquire plastic strains, were also mentioned as disadvantages by Sirota and Mal'tsev.

Pt vs Pt-(5-10) Au.   Thin film couples with this range of compositions were among those described in a patent by Hill (1963). The films are prepared by firing, after painting a substrate with liquid suspensions of the metals in an organic liquid. A thin conductive metal coating is formed as a result of the process.

TABLE 2-3

Comparison of Voltage-Temperature Data for Au vs Pt Thermocouples from Three Different Sources

Temperature ($^{\circ}$C)	Voltage (mV)		
	Roeser and Wensel (1941, p. 1293)	Sirota and Mal'tsev (1959)	Merryman and Kempter (1965)
100	0.78	- -	0.770
200	1.84	1.8392	1.834
300	3.14	3.1334	3.127
400	4.63	4.6221	4.623
500	6.29	6.2854	6.282
600	8.12	- -	8.115
700	10.13	- -	10.118
800	12.29	- -	12.288
900	14.61	- -	14.615
1000	17.09	- -	17.127

Iridium and Platinum

General Information. According to Hunt (1964) the thermoelectric investigations of iridium-platinum alloys date back to the work of P. G. Tait in 1872-1873. Tait studied alloys of platinum with up to 15 per cent iridium. The platinum-iridium metals have been little used, however, and the platinum-rhodium alloys later became almost universally preferred because of their superior performance.

Some interesting properties of iridium applicable to thermocouple usage were reviewed by Zysk (1964). Iridium has proved to be resistant to attack by lithium, sodium, potassium, bismuth, gallium, lead, silver, and gold at up to 200°C above the respective melting points of these elements. However, there is evidence of high temperature incompatibility with zirconium dioxide, often used as thermocouple insulation. In vacuum and in an argon atmosphere reaction occurs between iridium and lime-stabilized zirconium dioxide at 2260°C. Incompatibility has also been demonstrated with yttrium-oxide-stabilized zirconium dioxide and with thorium dioxide. No reaction has been noted at 2371°C with beryllium oxide or hafnium dioxide.

Pt-5Ir vs Pt. A thin film couple of this composition was described by Hill (1963), the preparation procedure being the same as for thin film Pt vs Pt-(5-10) Au couples.

Pt-10Ir vs Pt. This combination has had limited application, a disadvantage being that the iridium oxidizes at high temperatures, which changes the alloy composition. As a result the calibration falls with prolonged use. A voltage-temperature curve from 0 to 1600°C is shown in Figure 2-1 from Bennett (1961, p. 28) which is unusually linear through the entire range. The thermoelectric power, computed from this curve, is a little over 16 μV/°C.

Tests were conducted by Ihnat and Hagel (1960) in which this couple was exposed to the combustion products of fuel oil with 0.4 per cent sulphur at temperatures in the 1100-1320°C range. A drift was observed which exceeded 0.5 per cent of the initial output, measured at 1100°C, after 150 hours of exposure. Thermocouple failure occurred prior to 200 hours of exposure, apparently due to wire oxidation.

Pt-15Ir vs Pt. Using data from Zysk (1962), the output voltage at 1200°C is 22.3 mV and the thermoelectric power in the 1100°C region is slightly greater than 21 μV/°C. This and other types of thermocouples were evaluated by Boorman (1950) for performance changes caused by transition of a nuclear reactor from operating to shut-down conditions. The environment consisted of a neutron flux of 10^{12} n/cm^2-sec in a temperature range of 20 to 100°C. Other nuclear applications work was reviewed by Calkins and Schall (1960).

Ihnat and Hagel (1960) made an evaluation similar to that described for Pt-10Ir vs Pt and observed drift that exceeded 0.5 per cent of the 1100°C output after 50 hours, followed by failure before 150 hours exposure time was reached.

Pt-20Ir vs Pt. According to Hunt (1964) early studies of considerable precision were conducted by C. Barus, who was active at about the same time as Le Chatelier. Barus (1892) was particularly interested in Pt-20Ir vs Pt. He was aware of the importance of purity and homogeneity of thermocouple metals and his calibration work utilized the boiling point method, with mercury, zinc, and certain organic materials. However, the Pt-10Rh vs Pt couple recommended by Le Chatelier proved to be superior to the platinum-iridium thermocouples.

Ir vs Pt. A calibration curve is shown in Figure 2-1, plotted from data by Roeser and Wensel (1941, p. 1293). From this result the thermoelectric power in the 1000°C region can be found to be about 18 μV/°C. Troy and Steven (1950) conducted preliminary work with a number of materials and reported an output of 22 mV at 1500°C for Ir vs Pt. Walker and coworkers (1962) observed extensive drift when this combination was exposed to air at 1380°C for 120 hours, apparently caused by contamination of the platinum, either by iridium or by some impurity in the iridium.

Ir vs Pt-20Ir. A calibration was made by Troy and Steven (1950) over the temperature range of 982 to about 1900°C in an atmosphere of helium. The thermoelectric power appears to be in the order of 4 μV/°C in the 1400-1800°C range, decreasing as the temperature is increased. The voltage-temperature curve is included in Figure 2-1.

Ir vs Pt-30Ir.   This pair was also discussed by Troy and Steven, based on a calibration from about 1170 to 2000°C. The average thermoelectric power was near 5 μV/°C, but as can be seen from Figure 2-1, the voltage-temperature curve is nonlinear. Sanders (1958) included this pair in his review of a number of thermocouples and estimated a maximum error for application work of ±40°C compared with ±3°C for Pt-10Rh vs Pt.

Palladium and Platinum

General Information.   Palladium is known to be less inert than platinum in a number of environments, but there is far less thermoelectric information available in this respect than for platinum and rhodium. Zysk (1962) discussed the effect of environment on platinum group metal couples and indicated that in general palladium can be used in oxidizing atmospheres or in such inert gases as argon, helium, or nitrogen. Carbon dioxide appears to have no effect, but carbon monoxide exposure results in hardening, and the metal is known to absorb hydrogen. Palladium is attacked by nitric acid, by SO_2 above 800-1000°C, and by H_2S above 600°C. The latter atmosphere is not recommended for any thermocouples containing palladium.

Some oxidation of palladium occurs in air, beginning at about 700°C. Above about 875°C the oxide decomposes so that the metal surface remains bright. The rate of oxidation is low, however, so palladium can be annealed in air at 750-900°C. Caldwell (1962) reviewed annealing procedures and indicated that a 25-minute anneal at 750°C is satisfactory; a nitrogen atmosphere can be used to prevent oxidation, particularly at higher temperatures. The melting point of palladium is 1554°C, well below the 1772°C melting temperature of platinum. This, of course, limits its high temperature use.

Pt vs Pd.   This combination was employed as a thermocouple by A.C. Becquerel in 1826, five years after Seebeck announced his discovery of thermoelectricity. According to Zysk (1962), it was used by Becquerel to obtain a straight line voltage-temperature function to 300°C, which was later extrapolated to higher temperatures. The calibration data of Roeser and Wensel (1941, p. 1293) are plotted for the 0-1500°C range in Figure 2-2. The thermoelectric power near 1000°C is 20 μV/°C.

Ihnat (1959) and Ihnat and Hagel (1960) reported investigations of thermocouple drift in an oxidizing combustion atmosphere using fuel oil with 0.4 per cent sulphur; the test temperature was 1316°C. Four couples were tested, with somewhat greater variability between specimens than with other materials that were evaluated. Generally, drift was positive, although in one case a positive

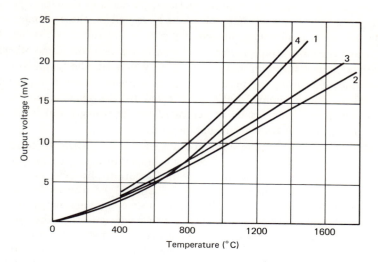

FIGURE 2-2   Platinum thermocouples with palladium, rhodium, and platinum-rhodium alloys.   1. Pt vs Pd, Roeser and Wensel (1941, p. 1293).   2. Pt-10Rh vs Pt, Shenker, et al. (1955).   3. Pt-13Rh vs Pt, Shenker, et al. (1955).   4. Rh vs Pt, Zysk (1962).

excursion from the initial calibration level reached a deviation of a little over 0.3 per cent at 150 hours and again at 475 hours, with a near return to zero net drift after 600 hours. The voltage measurements were made at 1093°C. One thermocouple showed continuous positive drift, reaching +0.5 per cent of the calibration voltage after about 210 hours. In general, nevertheless, the combination was considered to be suitable for 600 hours use without exceeding a +0.5 per cent output deviation.

Rhodium and Platinum

General Information.   Platinum and some of its alloys with rhodium have exceptionally good properties for thermocouple use. The thermoelectric power S of platinum-rhodium alloys relative to platinum is well behaved. The addition of rhodium to platinum increases the value of S, so that thermocouples consisting of platinum and of the alloy, or of two alloys of sufficiently different rhodium content, have an output that is adequate for measurement applications.

Some important physical properties of platinum and platinum-rhodium alloys are superior to those of other metals over a wide temperature range. The resulting advantages usually more than offset the importance of the larger thermoelectric powers available from many other combinations. The inertness of platinum and rhodium to chemical reactions, particularly oxidation processes, has played a major role in their selection. Nevertheless, these metals do have limitations, an example being the melting point of platinum, which lies below the range of extreme temperatures which can be reached with some other types of thermocouples. Platinum melts at $1772°C$, which sets the temperature limit for thermocouples using the pure element. Also, long-period use of platinum at temperatures above about $1400°C$ is affected by the phenomenon of grain growth, which not only results in decreased tensile strength but increases the likelihood of contamination at the grain boundaries.

Both platinum and platinum-rhodium alloys are sensitive to composition changes by means of diffusion at the high temperature junction, although the thermoelectric powers of the alloys are not as drastically affected as that of platinum alone. This property has provided one of the reasons for suggesting that all-alloy couples may be preferable to those employing pure platinum. In any case, diffusion of the constituents of either wire into the other at the hot junction interface results in a slowly changing composition. Thermocouple errors result which depend upon the initial compositions, the junction temperature, wire temperature gradients, and the cumulative high temperature exposure time.

Platinum, rhodium, and alloys of the two elements undergo surface evaporation when heated to high temperature in a vacuum. In air the weight loss of metal is due to surface oxidation followed by volatilization of the oxide. These weight loss processes are influenced by the environment, and specific conditions may give rather clear-cut loss-rate preference to either element. Fortunately, these processes are usually slow enough to permit use of the metals in oxidizing atmospheres at high temperatures.

Investigators have tended to be cautious about recommending thermoelectric use of these metals in vacuum, except for short periods of time. However, the high temperature vacuum weight loss does not appear to greatly affect voltage output, implying that in vacuum platinum-rhodium alloy constituents do not differ greatly in volatility. The possibility of contamination of one wire by volatile vapors from the other is a more important factor here. Long term vacuum applications are possible when each wire is separately contained in an insulating tube, or if twin bore tubing is used. The insulator must be a very pure material which is compatible with the wires at high temperatures, such as a very low impurity aluminum oxide.

Bare wires can be used without difficulty in atmospheres of commercially pure inert gases. Operation in helium, argon, or nitrogen is acceptable. Even carbon dioxide does not appear to react with these metals. Reducing gases such as carbon monoxide or hydrogen do not have adverse effects on bare wire couples which are self-supported in the high temperature region, provided that these atmospheres are free of contaminants, the wires are of high purity, and the wire surfaces themselves are contaminant-free.

The above statements regarding reducing atmosphere applications cannot be generally applied to couples supported by insulators, packed in insulation, or exposed to surface contact with other materials. Since most measurement requirements include one or more of the above conditions, platinum-rhodium bearing thermocouples are not normally recommended for reducing atmosphere applications.

One of the principal difficulties encountered in reducing atmospheres is the reduction of silicon dioxide, which is often present in trace quantities in even the best quality metal-oxide insulators, or in other contacting materials. When reduction occurs, metallic silicon is formed, which diffuses into the platinum or platinum alloy; this lowers the melting point, causes embrittlement, and affects the thermoelectric power. If traces of carbon and sulphur are also present, a rapid atmospheric attack on the platinum results, even when the silicon dioxide-bearing material is not in direct contact with the metal. It also appears, however, that in the complete absence of silicon the reducing atmosphere does not affect performance in the presence of carbon and sulphur.

The use of insulators or crushed insulation in vacuum, inert atmospheres, or oxidizing atmospheres is often recommended and is common practice. Considerable work on the compatibility of insulation materials with platinum and platinum-rhodium alloys has been done with thermocouple metal pairs and is discussed with other results under the appropriate thermocouple. Results presented by Chaussain (1952), Ehringer (1954), Rudnitskii and Tyurin (1956), and Walker, Ewing, and Miller (1962; 1965, p. 601) give much information on insulation materials. The sensitivity of these materials to impurities leads to considerable variation in behavior, however, and it is not surprising to find some lack of agreement among the various investigators. In fact, the important effects of iron impurities on thermocouple drift were not known until the work of Walker, et al.; hence much of the earlier information may be rather limited in usefulness. It is now well established that iron, rather than silicon, is more detrimental to thermocouple performance in oxidizing or inert atmospheres, or in vacuum.

Darling and Selman (1971) discussed the compatibility of the platinum metals with insulating materials such as the oxides of magnesium, silicon, and zirconium. At high temperatures metal reactions with oxide insulation may take place under either inert or reducing conditions and with or without mechanical contact, since most such reactions can proceed via the vapor phase.

Pt-10Rh vs Pt.   An outline of the early history of this thermocouple was given by Hunt (1964). It was first investigated by H. Le Chatelier in 1885 while making studies concerned with cement manufacture. He selected Pt-10Rh vs Pt after investigating a number of metals, and also recommended that calibration be in terms of the fixed melting or boiling points of certain pure substances. This has since become a highly regarded technique for accurate work. Finally, Le Chatelier spent time and effort over a five-year period in developing a pyrometer and arranging for its manufacture, so that by 1890 thermoelectric thermometry was practical for industrial applications as well as for the laboratory.

According to Darling (1961), L. Holborn and W. Wien demonstrated the low drift characteristics of this couple during work in the 1892-1899 period, making a favorable comparison with the performance of the platinum resistance thermometer. Sosman (1910) later reported results for the 0 to 1755°C range,[3] and he prepared the first thermocouple reference table for this combination. In the high temperature range this couple became the most accurate measurement means available. A number of revised and improved reference tables were introduced in later years, the most widely used work being that of Shenker, et al. (1955) in NBS Circular 561. The most recent table is that of Bedford, Ma, et al. (1971).

The International Temperature Scale of 1927 defined temperature in the 660 to 1063°C range in terms of the voltage of a Pt-10Rh vs Pt couple. A couple which met certain composition and purity requirements and which was calibrated at the freezing points of antimony, silver, and gold determined the temperature at intermediate points. Revisions of the scale in 1948 and 1960, which were described by Stimson (1961), and a later revision in 1968, retained the thermocouple method of temperature determination for what is now called the International Practical Temperature Scale (IPTS). On the IPTS, the output of the Pt-10Rh vs Pt couple is used for temperature definition in the interval between 630.74°C and the gold point at 1064.43°C. According to Barber (1969) temperature is still defined by the formula $V = a + bT + cT^2$, where V is the thermocouple voltage. The constants a, b, and c are determined by the values of V at 630.74 ±0.2°C, the silver point, and the gold point. The value of V at 630.74°C is determined by measuring that temperature with a standard platinum resistance thermometer.

Jones (1969) converted British Standard 1826, which is the equivalent of NBS Circular 561 used in England and elsewhere, to the IPTS 1968. Later, new reference tables were prepared by scientists at the National Research Council of Canada, the National Physical Laboratory in England, and the National Bureau of Standards, as the result of a joint program. The work was reported by Bedford, Ma, et al. (1971).

Bedford (1971) described the results of a large number of calibrations and intercomparisons, from which it was shown that the IPTS of 1968 is specified to within ±0.1°C between 630.74 and 1064.43°C, from its definition in terms of properly selected Pt-10Rh vs Pt thermocouples. These, and also Pt-13Rh vs Pt couples, differ from the scale by about ±0.1°C at most, and probably are within ±0.05°C.

The materials used in the standard thermocouple and for couples required in high accuracy applications are carefully prescribed. For IPTS definition, platinum purity is maintained by requiring that the resistance of the wire at $100°C$, and at $0°C$, expressed as the ratio R_{100}/R_0, must be not less than 1.3920. The thermocouple voltage of the alloy, paired with the pure platinum, must meet very specific requirements. These have been discussed, and in some cases criticized, as efforts have been made to further improve the practical temperature scale. Estimates of subsequent thermocouple accuracy and other pertinent description and discussion of this subject include work by Stimson (1961), Jones (1968), Bentley (1969), Bedford (1971), and Gray and Finch (1971). The latest reference tables, which were described by Bedford, Ma, et al. (1971), are based on alloy wires that are as close as possible in composition to Pt-10Rh.

The Pt-10Rh vs Pt thermocouple can be used for precise measurements both above and below the limits of the 630.74 to 1064.43°C range. For example, Roeser and Lonberger (1958) stated that by using the zinc point together with the antimony and gold points, the temperature range can be extended down to 400°C. Agreement with the International Temperature Scale of 1927 was to within 0.5°C between 400 and 1100°C. There may be some small change in the agreement when compared with the present IPTS scale.

Measurements of temperature and of small temperature differences in the range 20 to 1200°C were made by Lyusternik (1963) after calibrating by the comparison method and developing equations for the voltage-temperature characteristic. The equations agreed with the calibration data to better than 0.1 per cent. A recalibration made after three years of use up to 1200°C showed no appreciable change from the original results.

Voltage at a temperature of 1000°C was compared with results of earlier investigators; some of this information is included here in Table 2-4. Thermoelectric power is also shown. Additional values of the latter quantity at other temperatures are given in Table 2-5. It was pointed out that the calibration results supported an earlier opinion of Roeser and Wensel (1941, p. 284) that difference curves between calibrations of individual couples are almost linear functions of temperature.

Methods of extending the calibration range over intervals as great as 0 to 1500°C have been outlined by Finch (1962), using the procedure of taking additional fixed point calibrations, and by comparison with a standard platinum resistance thermometer. Extrapolation is used above the gold point, with checks made by means of an optical pyrometer. Jones (1971) discussed uncertainties introduced by different interpolation methods and also stated that calibration uncertainty in using thermocouple intercomparison technique is equal to or less than the uncertainties associated with fixed point calibrations. Kostkowski and Burns (1963) indicated that a repeatability of 0.1°C is attainable up to at least 1100°C for short time periods, using carefully handled, well annealed, reference grade wires. According to Hall (1966) this figure represents about the best repeatability that can be attained, taking all precautions to minimize errors.

TABLE 2-4
Voltage and Thermoelectric Power of Pt-10Rh vs Pt at 1000°C Compared from Different Sources

	Roeser and Wensel (1933)	Tanaka and Okada (1937)		Lyusternik (1963)
Voltage, μV	9569.3	9616.8,	9599.9	9578.8
Thermoelectric power, μV/°C	11.530	11.597,	11.565	11.541

TABLE 2-5
Thermoelectric Power of Pt-10Rh vs Pt in the 0-1200°C Range[a]

Temperature (°C)	S (μV/°C)
0	5.60
200	8.43
400	9.52
600	10.22
800	10.90
1000	11.54
1200	12.09

[a]Lyusternik (1963).

Wire for Pt-10Rh vs Pt thermocouples is commercially available in either standard or reference grade quality; the latter type is sometimes called premium or special grade. Such wire meets the voltage-temperature characteristics of the current reference tables to within certain limits of error or tolerance. Starr and Wang (1969) stated that wires are available with a tolerance of ±0.5 per cent (standard grade) and ±0.25 per cent (reference grade) for the temperature range above approximately 600°C. Some suppliers offer a reference grade tolerance of ±0.1 per cent above 600°C and attempt to keep the voltage deviation from the tables in the positive direction. Caldwell (1962) pointed out that the upper limits of temperature at which the percentage tolerances apply vary somewhat depending on the manufacturer and the tolerance selected. Values range from 1316 to 1482°C. Below about 600°C the tolerance is usually given in degrees, typical values being ±2.8°C and ±1.4°C for standard and reference grades, respectively.

The performance of couples made from these wires can be quickly determined by the user by means of a one point check at a suitable temperature; this can be done since the difference curve for the deviation of any given thermocouple from the tabular values is often practically a straight line.

Temperature measurements can be made for brief periods, perhaps with reduced accuracy, at temperatures well above 1482°C. The upper limit is the melting point of the platinum, 1772°C. Finch (1962) recommended a one point check for possible calibration changes after using a couple at temperatures as high as 1600°C, however. More recently, Bedford (1969) suggested that these couples have been so improved in manufacture that they are now more stable and accurate at 1400°C than they once were at 1100°C, and that short time use above 1600°C is not uncommon.

Bedford also pointed out that low temperature use is now feasible for applications where it would be advantageous to use a single couple over a wide range of temperature. An example given was that of thermal conductivity measurement where the temperature might range from below –50°C to 1100°C. The thermoelectric power of Pt-10Rh vs Pt is low below 0°C, but accurate measurements are still possible with instruments now available. The lower end of the temperature measurement range is still given as 0°C for general applications.

A voltage-temperature curve is shown in Figure 2-2, where it can be compared with some other platinum-rhodium combinations. Suppliers and interested organizations designate this couple as the Type S thermocouple. Extension wires of copper and copper-nickel alloy are available from most suppliers that offer the noble metal wires or cables.

The noble metal wires are usually annealed before shipping. However, unless very carefully handled both during preparation of the wire for shipment and during preparation of the thermocouples, there may be sufficient cold working to produce measurement errors. Such errors occur if an inhomogeneity caused by an induced strain in the wire is located in a temperature gradient during use.

Errors caused by cold working are very difficult to analyze or predict, since the degree of cold working, the temperature gradient, and other factors all contribute to any output deviation that may exist. Experiments made by Kollie and Graves (1966) described the performances of a cold-worked thermocouple and an annealed unit, and showed that the former was in error by –8°C in the 200 to 1000°C region. The annealed couple was in error by –1°C under similar conditions. A type of hysteresis effect was also observed in the cold-worked couple, due to annealing which took place during calibration. It was concluded that cold working can seriously alter the thermoelectric voltage.

Other results by Pollock and Finch (1962) also indicated that changes are produced by cold working, and showed that in most instances the maximum voltage deviation occurred immediately after cold working with no significant evidence of rapid recovery afterwards. Because of the possibility of cold work induced errors, annealing just before use is desirable when the best possible performance is required.

A number of annealing schedules have been recommended for this type of thermocouple. According to the American Society for Testing and Materials (1970, p. 108) it is good procedure to electrically self-heat the thermocouple wires in air while they are loosely hanging between two binding posts. The National Bureau of Standards has adopted the procedure of annealing for one hour at 1450°C. Such a procedure is probably adequate for most general applications. McLaren and Murdock (1971) demonstrated, however, that an additional 16-hour anneal at 450°C is desirable when the very best performance is required. These investigators, in fact, recommended a 1300-1450°C anneal of one hour for the platinum wire, and ten hours for the alloy, then a one-hour anneal for the assembled thermocouple, followed by the 450°C anneal. Additional treatment was also recommended for removal of oxide film and for strain relief between periods of use. Oxide stripping[4] was accomplished by a ten-minute treatment at 1300°C. Following this and each high temperature application exposure the 16-hour, 450°C anneal was advised.

Bare wire Pt-10Rh vs Pt couples can be used without protection in oxidizing or inert atmospheres. A reducing atmosphere is not generally recommended, particularly above about 900°C; such usage may be possible under the very limited conditions outlined earlier in the general information under Pt-Rh alloys, and given in more detail in following paragraphs. Short time operation in vacuum is feasible. Hendricks and McElroy (1966) found that long period vacuum operation is possible for couples within two-hole aluminum oxide sheaths. Tests by Szaniszlo (1969) with Pt-13Rh vs Pt couples in such sheaths showed an average drift of slightly over 0.2 per cent of the true Celsius temperature after 3700 hours exposure at 1257°C and 3×10^{-8} torr.

Estimates based on experience were made by Kostkowski (1960) and Kostkowski and Burns (1963) for the amount of drift to be expected in oxidizing atmospheres. After ten hours at 1100 or 1200°C, errors of 0.1 or 0.2°C were estimated. After 100 hours at 1200°C, a figure of 2°C was given. A figure of 1°C was stated for ten hours operation at 1500°C.

Glawe and Shepard (1954) found that exhaust gases of burning propane, gasoline, or JP-4 fuel, directed at bare wire hot junctions, produced negligible calibration changes. This same result occurred regardless of whether the gases were rich, lean, dirty, or clean, and regardless of coating formation on the wires. Calibration changes were always less than 2.8°C at 982°C. These tests were conducted using Pt-13Rh vs Pt couples; the results are also applicable to Pt-10Rh vs Pt.

According to Homewood (1941) and Brenner (1941), bare wire couples operated for periods of time in air, with steady state temperature conditions, develop a distinctive dark surface coating on the portions of the alloy wires which are in the 400-600°C temperature region. This discoloration is said to be due to the formation of rhodium oxide. The length of the discolored zone reaches a maximum value which remains relatively constant for long periods. Homogeneity tests by Homewood showed that the blackened section was less positive than normal relative to platinum, indicating a loss of rhodium, and Brenner concurred with this explanation.

McLaren and Murdock (1971) reported formation of a similar black surface layer on the alloy wire in the 600 to 850°C zone. Evidence was cited that for this type of air exposure in the 400-850°C temperature range the black discoloration is due to the presence of rhodium sesquioxide, Rh_2O_3. This oxide is stable to about 1100°C and can be removed by exposure to higher temperatures.

It seems well established that in air it is the rhodium which oxidizes most rapidly, particularly where the wires are led out through a temperature gradient from a heat source at high temperature. Nevertheless, for many other conditions such as furnace interiors and sealed chambers where other types of atmosphere may exist, there is disagreement among various investigators as to whether it is platinum oxide or rhodium oxide that volatilizes more rapidly and thereby changes the composition of Pt-10Rh or Pt-13Rh alloys. There is also a question of which element volatilizes more rapidly in an inert atmosphere or vacuum. In marginally oxidizing atmospheres the possibility exists of competing processes of oxidation and direct vaporization. As stated earlier, Homewood and Brenner considered rhodium to be the more volatile. Jewell and Knowles (1951) and Svec (1952) found comparable results. On the other hand, McQuillan (1949) reported that platinum has a higher volatilization rate. Zysk (1962) discussed this problem and included several other sources supporting McQuillan. Walker and coworkers (1962), nevertheless, found no significant evidence of preferential volatization of either element during work at temperatures of 800-1700°C.

Hill and Albert (1963) reported studies of platinum and rhodium sample weight losses in several atmospheres. In general, platinum weight loss was either higher or about the same as that of rhodium. A somewhat higher weight loss for rhodium at 1700°C in argon was shown, however. In an inert atmosphere or in a vacuum the metals vaporize without prior oxide formation. It is evident from this that in mildly oxidizing atmospheres both oxidation and vaporization may be present as competing processes. Other conditions such as the flow velocity of gas past the wires can also affect the results. Since a number of factors are involved, the discrepancies reported by various investigators may be caused by different processes predominating in each case.

Dallman (1964) described a coating process for eliminating volatilization. Each thermocouple wire was coated with a layer of aluminum oxide, and this prevented the volatilization of platinum and rhodium in flowing air. Even when volatilization does not occur, however, diffusion of constituents from one wire to the other can take place at the thermocouple junction. In this situation there is migration of rhodium from the alloy wire to the platinum during long periods at high temperatures, resulting in contamination of the pure metal.

At relatively low temperatures where volatilization and diffusion are not important, platinum-rhodium thermocouples can be used for thousands of hours in some cases. Darling, et al. (1970, 1971) reported that a Pt-13Rh vs Pt couple used for 40,000 hours at 625°C drifted by −2°C. In this case the drift was caused by internal oxidation of rhodium within the alloy conductor in the section of wire that was within the 400 to 600°C temperature region.

Some atmospheres affect thermocouple performance at high temperatures when insulators and other materials are present. In most such circumstances the use of Pt-10Rh vs Pt and other platinum-rhodium thermocouples is not recommended in reducing atmospheres. There are exceptional cases, however, as shown by several investigations for the British steel industry; it was determined that contamination of platinum-rhodium thermocouples takes place in the presence of a reducing atmosphere plus silica, sulphur, and carbon. This work, introduced by the Liquid Steel Temperature Subcommittee[5] (1947), consisted of reports by Land (1947), Reeve and Howard (1947), Goldschmidt and Land (1947), Manterfield (1947), Chaston, et al. (1947), and Jewell (1947). The process of contamination was found to be complex. In brief, the action of the reducing atmosphere is to convert silica to metallic silicon which then combines with sulphur in the presence of carbon monoxide to form a vapor of SiS_2. This vapor then attacks the platinum metals and silicides of platinum are formed.

McQuillan (1949) made a study of this same problem and concluded that if special precautions were made to eliminate carbon and sulphur, a hydrogen atmosphere at temperatures up to about 1600°C could be used with silica present. She also concluded that couples could be used with that same atmosphere and temperature in the presence of carbon, beryllium oxide, or aluminum oxide, provided that no silica-bearing materials were present.

Zysk (1962) summarized practical steps to follow when preparing thermocouples, so that at least some of the potential contaminants can be removed. Since carbon, and often sulphur, are evolved during the decomposition of greases and oils, care should be taken to remove all traces of lubricating oils or greases during preparation. Even minute amounts of grease from the fingers may cause contamination, so a nitric acid wash of the bare wires followed by an alcohol rinse is often recommended. Bennett (1958) also emphasized this aspect of the contamination problem and described the use of thin leather gloves for handling the wires.

Zysk also described thermocouple failure due to silicon contamination of the wires, which is characterized in the initial stage by an erratic thermocouple output. Fractures in wires, caused by platinum silicides, generally present a melted appearance. The surface of the wire in the vicinity of a break frequently contains a number of glazed areas.

The studies of silicon contamination of platinum and platinum-rhodium alloys described previously were not sufficient to explain the drift often observed when insulator-sheathed or insulation-packed wires are used in inert or oxidizing atmospheres, or in vacuum. Walker, Ewing, and Miller (1962; 1965, p. 601) showed that under these conditions the usual source of contamination is iron rather than silicon. These investigators reported that the amount of iron in the best insulators then available was less than 0.04 per cent by weight, and that this was still sufficient to produce drift during periods of time at high temperatures.

Drift tests in air revealed calibration changes of -0.6 to -1.4°C, after 240 hours exposure at 1380°C in air. Drift was less at lower temperatures and was zero at 1000°C. Other thermocouple combinations, with platinum-rhodium alloys used for both conductors, were less sensitive to contamination than Pt-10Rh vs Pt. Experiments in air, argon, and vacuum with Pt-10Rh vs Pt and the best performing aluminum oxide insulators showed no calibration change at 860°C following 120 hours exposure at temperatures up to 1380°C. With a 1600°C exposure the results were -1, -11, and -4 Celsius degrees, measured at 860°C, for air, argon, and vacuum, respectively. Results were less satisfactory with insulators of lesser purity. Drift was independent of wire size in air. In argon and vacuum drift was less with larger wires. Wires of 0.51 mm diameter were used for determining the figures given here.

Other stability and drift results were reported by Bostwick (1962) for thermocouples in aluminum oxide insulators. The size of the wires, the type of thermocouple junction, and the method of junction preparation all affected the performance of the thermocouples. It was recommended that the use of hydrogen arc welded junctions be avoided.

Hendricks and McElroy (1966) used two-hole aluminum oxide tubing which had been pretreated to remove as much iron as possible. Thermocouples in these insulators were exposed to high temperatures in a vacuum of 2×10^{-7} torr. Drift was within the limits of ±5°C for 352 hours at 1300°C and within ±8°C during an additional 1012 hours at a temperature of 1450°C. Szaniszlo (1969) used two-hole 99.5 per cent purity aluminum oxide insulators with Pt-13Rh vs Pt wires and observed drift that averaged -2.8°C, measured at 1257°C, after 3700 hours at that temperature and in a vacuum of 3×10^{-8} torr. Glawe and Szaniszlo (1971) reported additional work in air, argon, and vacuum for periods as long as 10,000 hours. This work was done with Pt-13Rh vs Pt but the results should also be applicable to Pt-10Rh vs Pt.

Greenberg (1971) described a thermocouple used in a glass industry application which showed low drift during operation at 1260°C. The wires were protected by aluminum oxide powder within a double bore aluminum oxide insulator, and after a three-year operating period at 1260°C a drift of -12°C was found at a calibration temperature of 1200°C.

Darling and coworkers (1970, 1971) recommended the use of magnesium oxide as described in their series of journal articles and conference paper. This oxide is particularly suitable when exceptionally small amounts of oxygen are present; under this condition it was found that other refractories reacted with platinum and its alloys. Zysk (1962) described other types of insulators that have been widely used with Pt-10Rh vs Pt. Below 1000°C, quartz, mullite, or porcelain is used. Synthetic mullite has been used up to 1450°C in oxidizing atmospheres; caution was recommended concerning reducing atmospheres because of its high silica content. Quartz has also been used at higher temperatures, but usually for short, quick-immersion conditions where its superior resistance to thermal shock is desirable. Beryllium oxide, zirconium dioxide, and thorium dioxide have all been used for some applications. Darling, et al., described reactivity with the latter oxide unless the amount of phosphorus present as an impurity was less than 75 ppm. Greenberg (1971) discussed use of various insulation materials as well as several types of sheaths, with emphasis on glass industry applications.

A mixed oxides coating for bare wire thermocouples was described by Kent (1970); this prevented catalytic action, which is a problem with platinum metal thermocouples in some atmospheres. Coatings of yttrium oxide with 10 to 15 per cent beryllium oxide had a fusing temperature below the melting point of platinum and at the same time could be used for protection to above 1600°C. No reactivity between the insulation and the thermocouple was observed, and there was no drift within 0.5 and 1.0 per cent experimental error limits for 15 and 20-hour tests in two types of flames. The mixed oxides coating was said to have a high electrical resistance at 1600°C.

There is a lack of quantitative information on the drift of Pt-10Rh vs Pt or Pt-13Rh vs Pt when in contact with materials other than aluminum oxide and magnesium oxide. For comparative purposes, some of the results of Chaussain (1952) are included for wires embedded in fine powders of various materials, and these are given in Table 2-6. These results were for exposure in an atmosphere of air at 1300°C and with relatively pure materials, but the results may have been influenced by the presence of iron or other contaminants.

Direct contact of Pt-10Rh vs Pt thermocouple wires with a metal may result in contamination caused by diffusion of atoms from the metal into the noble metals or vice versa. Jensen, Klebanoff, and Haas (1964) reported drift caused by diffusion, on the basis of experiments with couples spot welded to nickel and operated at 950°C in a vacuum. A decline of zero to 3 per cent in voltage was observed during the first day of tests with 12 couples. After some 24 days, the decline ranged from about 18 to 30 per cent. Comparative tests using couples spot welded to platinum showed no such changes; it was concluded that contamination from the nickel must have affected thermocouple performance.

TABLE 2-6

Drift of Pt-10Rh vs Pt Thermocouples Embedded in Powdered Oxides and Heated to 1300°C

Material	Drift in Celsius Degrees After the Time Shown	
	12 Hours[a]	20 Hours[b]
ThO$_2$	-2	- 3.5
MgO	-5	- 7.3
ZrO$_2$	- -	- 8.5
CaO	- -	-10.7
Al$_2$O$_3$	-7	-11
SiO$_2$	-9.5	-15.4

[a] From tabulated data.
[b] Estimated from graphs of drift versus time.

A thermocouple may be unstable, once contaminated by a base metal, even if it is afterwards operated in a way which prevents further contamination. Fairchild and Schmitt (1922) compared the drift of a Pt-10Rh alloy with about 0.34 per cent iron with that of an iron-free alloy. After 18 hours at 1250°C, the iron-bearing alloy had drifted -9°C compared with -2°C for the iron-free metal. After six hours at 1450°C, the drifts were -11°C and -3°C for the iron-bearing and iron-free alloys, respectively. In each case, the drift was measured at 1200°C.

There is evidence that some wire impurities have a stabilizing effect on thermocouple drift when the amount of impurity and of further contamination during use is small. Examples of this condition occur with standard grade couples, compared with premium or special grade couples which are exposed to identical conditions, when the latter show a greater drift rate. Presumably the purer premium thermocouples are more sensitive to initial contamination. Nielson (1970) demonstrated this condition with Pt-13Rh vs Pt and showed that in one case the purer type of thermocouple drifted about twice as fast as the standard grade couple.

High purity carbon is said to have no effect on platinum metals, and couples exposed to carbon have remained unaffected in at least some cases. In other situations drift has been reported; negative drift, embrittlement, and failure conditions were described by Carroll and Reagan (1966) at temperatures above 900°C and with an inert atmosphere. Participation of carbon in the four-factor contamination mechanism including silica, sulphur, and a reducing atmosphere was discussed earlier. Since minute impurities are known to play an important role in carbon exposure, it is probably best to avoid carbon or graphite contact where possible.

Protection tubes, wells, or sheaths have been commonly used to isolate the Pt-10Rh vs Pt thermocouple from undesirable environments. One of the earliest and most widespread applications of this type was for liquid metal temperature measurements in the steel industry. Hatfield (1917) described this type of use in the World War I period. The Pt-10Rh vs Pt couple has never been fully satisfactory for the application, however, because the required temperatures are near its upper limit. Nevertheless, protection methods were successful enough for Clark (1946) to describe this couple as the standard device for measuring liquid steel temperature, against which all other measurement methods were compared and judged.

The development of swaged and drawn metal sheathed cables with high temperature insulation greatly broadened the field of applications for thermocouples, since to some degree the sheath can be tailored for a particular environment while isolating the thermocouple wires from outside contamination. Pt-10Rh vs Pt thermocouples have been used with a number of metal oxide insulation materials as well as a variety of sheath metals.

The behavior of Pt-10Rh vs Pt and Pt-13Rh vs Pt with noble metal sheaths was described by Freeman (1962), Black and Wilken (1963), Zysk (1964), and Hancock (1967). More recently, Selman (1971) and Selman and Rushforth (1971) reported that the drift which sometimes occurs in sheathed thermocouples at temperatures above 1300°C is attributable to the presence within the sheath of mixed platinum and rhodium oxide vapors from which the platinum conductor can become contaminated by rhodium. This problem was overcome by the use of hermetically sealed sheaths, which contained an inert atmosphere rather than entrapped air within the insulation; this arrangement resulted in performance comparable with bare wire thermocouples.

Inspection of unsealed Pt-10Rh sheathed couples with magnesium oxide insulation revealed reddening of the insulation after long heating. Rhodium contamination resulted in negative drift of as much as 200°C after 1400 hours at 1450°C. Insulation quality was also important and drift was more pronounced when the insulation contained absorbed water vapor and showed lower than normal insulation resistance. Tentative drift characteristics were shown for sealed sheath couples which were argon filled and magnesium oxide insulated. It was indicated that the hermetically sealed couples would drift by about -4 to -10°C after 500 hours at 1450°C with an atmosphere of air around the sheath.

Some choices of base metal sheaths result in problems. The use of molybdenum sheaths with beryllium oxide or magnesium oxide, and niobium sheaths with magnesium oxide, resulted in various instability effects ranging from about 1 to 22 per cent at temperatures above 1200°C; these results were obtained by Black and Wilken (1963). Examination revealed that a migration of elements had taken place; sheath material was found in the insulation, and metal from the reduced oxide insulation was found in the thermocouple wires.

Zysk (1964) described work where tantalum-sheathed aluminum oxide and magnesium oxide insulated couples were tested at 1600°C in argon, and which revealed instabilities and drift. An aluminum oxide insulated couple changed by decreasing 27°C in indicated temperature after 95 minutes. A magnesium oxide insulated unit showed poorer performance by changing −61°C in 60 minutes. Work by Hendricks and McElroy (1966) with tantalum and niobium sheaths and magnesium oxide insulation showed large negative thermocouple drift during 143 hours at about 1200°C. Drifts of −56 and −64°C were observed with tantalum and niobium sheaths, respectively. Observations by Moore and McElroy (1966) indicated that Inconel-sheathed aluminum oxide insulated couples would not perform satisfactorily under at least some conditions. It is evident from all of these results that reactions between the thermocouple metals, the oxide insulation, and the sheath or other nearby metal occurred in situations where no reaction would have been anticipated, based on earlier experience with the thermocouple and the insulation alone.

Both bare wire and sheathed Pt-10Rh vs Pt couples have been used to some degree for nuclear applications, particularly in situations where reactor temperatures were expected to somewhat exceed the temperature limitations of Chromel-Alumel thermocouples. The good temperature stability and drift characteristics of Pt-10Rh vs Pt often encourage long term use in spite of the error produced by transmutation of the thermocouple elements. The latter effect is primarily due to the high thermal neutron cross section of rhodium. Neutron bombardment transmutes some of the rhodium to palladium, which changes the thermocouple composition, and hence its thermoelectric properties. A small percentage of platinum is also transmuted, first to gold and finally to mercury.

Browning and Miller (1962) made calculations of the change of composition after ten years of irradiation in a neutron flux of 10^{14} n/cm^2 sec, and showed that most of the rhodium in the Pt-10Rh alloy would be converted to palladium. Computations by Ehringer and fellow investigators (1966) for integrated fluxes of 10^{20} and 10^{21} n/cm^2 predicted the presence of palladium in concentrations exceeding 0.2 per cent and of about 1 per cent, respectively. They also indicated that there would be sufficient platinum transmuted to be noticeable, and concluded that the Pt-10Rh vs Pt couple cannot be regarded as reliable after neutron exposures exceeding 10^{20} n/cm^2.

Experimental work consisting of radiation exposure of thermocouples and qualitative or comparative examination of performance was reported by Trauger (1960) and by Kelly (1960). Carroll and Reagan (1966) observed no significant effects for an integrated flux of 3.6 x 10^{20} n/cm^2 of thermal neutrons and 2.8 x 10^{20} n/cm^2 fast neutrons. Changes exceeding ±5 per cent would have been detected. Hancock (1967) concluded that Pt-13Rh vs Pt couples would probably show an error not greater than 5 per cent per 10^{20} n/cm^2 for thermal neutrons. Pt-10Rh vs Pt performance would be similar.

Results by Ross (1961), Levy, et al. (1962), and Kelly, et al. (1962), based on post-irradiation calibration of exposed couples, indicated decreased thermocouple voltages due to transmutation errors that ranged from –0.8 to –3 per cent for integrated neutron flux in the 1.6×10^{20} to 4.2×10^{20} n/cm^2 category. At 3.8×10^{21} n/cm^2, for neutron energies exceeding 0.18 meV, an error of –14 per cent was reported by Helm (1967).

Dau and coworkers (1968) reviewed work by Madsen (1951) and other sources concerning effects of nuclear heating and the short term reversible changes in calibration which are proportional to the flux level. It was concluded that dose rate effects on calibration are less than 1 per cent for flux of less than 10^{13} n/cm^2 sec.

The behavior of Pt-10Rh vs Pt at high pressures was studied by Bundy (1961), Hanneman and Strong (1965, 1966), Peters and Ryan (1966), and Getting and Kennedy (1970). Studies in a combined environment of high pressure and high temperature are difficult and there are some discrepancies among the results of the various investigators.

This type of thermocouple was found to indicate a temperature lower than the true temperature when the hot junction and the heated wire lengths were at a high pressure. For example, additive correction curves by Getting and Kennedy indicated that a corrective voltage of +180 μV is required when the pressure is 35 kB and the hot junction temperature is 1000°C. This is equivalent to about 15.5 Celsius degrees. The curves by Getting and Kennedy were based on experimental results up to 35 kB and 1000°C and were shown with extrapolations up to 50 kB and 2000°C, where the correction is +28°C. The uncertainty of the corrections was said to be ±(10% + 10 μV) for the experimental portions of the curves, which leads to an uncertainty of ±0.2 per cent in the measured temperature after correction. The latter figure is small because the correction itself is small. A detailed discussion of the use of the correction curves as well as of the experimental work itself was included.

Pt-13Rh vs Pt.   This combination has an unusual history, which began when Fairchild and Schmitt (1922) were investigating an inconsistent drift of various Pt-10Rh vs Pt couples. It was discovered that couples of English manufacture contained an impurity content of 0.34 per cent iron. This was later found to originate with the rhodium used to prepare the Pt-10Rh alloy wire. Presence of the impurity led to a noticeable increase in the thermoelectric power and resulted in a tendency for drift at high temperatures which was more pronounced than that of iron-free couples.

The manufacturer subsequently presented iron-free thermocouples which compared well with those from other sources; nevertheless, the iron-bearing couples were already in widespread use, particularly in England. Therefore, a new iron-free couple was produced for general replacement of the latter, using 13 per cent rhodium in the alloy in order to match the existing voltage-temperature characteristics of the iron-bearing alloy. As Zysk (1962) pointed out, this expedient has become self-perpetuating.

The voltages of the Pt-13Rh vs Pt and the Pt-10Rh vs Pt thermocouples are similar over the full range of temperature, as shown in Figure 2-2. This is true in general of all the other properties of the two thermocouples. Except for the respective temperature reference tables, and where specific exceptions are made, the discussion of one applies also to the other. Pt-13Rh vs Pt is designated as the Type R thermocouple.

The establishment of reference tables for this combination in the United States and Canada began with the investigation by Roeser and Wensel (1933) and, with later revisions, led to the work of Shenker, et al. (1955). The latter table was used until the advent of the International Practical Temperature Scale of 1968, with new tables soon after by Bedford, Ma, et al. (1971). Other precision calibration work was done by Barber (1943, 1950); the resulting tables were said by Bedford (1969) to be widely used in Europe.

The behavior of this couple was the subject of work by a number of investigators which was presented in a report of the Liquid Steel Temperature Subcommittee of the Iron and Steel Institute in England (1947). The report discussed the contamination process described earlier under Pt-10Rh vs Pt, which involved the four contributing factors consisting of a reducing atmosphere, silicon dioxide, sulphur, and carbon. Although the work was done for Pt-13Rh vs Pt in a special application, the results are generally applicable to thermocouples utilizing platinum and rhodium.

Possible drift due to interdiffusion between the two thermocouple metals at the hot junction was analyzed by Mortlock (1958) for Pt-13Rh vs Pt, and it was determined that in a temperature gradient of $10°C/cm$ the couple may be in error by about $1.3°C$ in the normal operating temperature region, following 100 days exposure at $1500°C$. This result applies to a conventional junction at the joined ends of two parallel wires. At lower temperatures where diffusion is not so important Darling, et al. (1970, 1971) described internal oxidation of rhodium in the alloy conductor, which took place over a 40,000-hour period when a couple was operated at $625°C$. Oxidation occurred in the length of wire where the temperature was in the 400-600° region.

Tests made by Szaniszlo (1969) in a vacuum of 3×10^{-8} torr with 99.5 per cent pure aluminum oxide insulators showed very small negative drift for couples exposed 3700 hours at $1257°C$. The average drift of six thermocouples was $-2.8°C$ or a little over 0.2 per cent of $1257°C$. The spread in the data was $1.6°C$. The ratio of the wire resistance at 300 K to that at 4.2 K was also measured during the tests and also showed a slight drift, which matched the thermoelectric drift, and indicated increased contamination of the platinum wires. An emission spectrographic analysis of wire impurities was made as well as a study of test-induced grain structure changes. It was found that the concentration of wire impurities had increased, with iron having the greatest increase at the hot junction and the largest concentration gradient near the junction. Additional work in vacuum with similar results was reported by Glawe (1970), and Glawe and Szaniszlo (1971).

The latter papers also reported drift tests in air and in argon with thermocouples in aluminum oxide insulators of not less than 99.5 per cent purity. In air the maximum drift was -5°C after 10,000 hours at 1327°C. Final drift in argon over a like period of time ranged from -18.8 to -27.6°C for three thermocouples. Drifts in both argon and vacuum were considerably higher than in air, which led to the conclusion that impurities which in air were converted to harmless oxides remained to contaminate the thermocouples when the other environments were used.

Wire size was a factor in determining drift in argon and vacuum. Wires of 0.51 mm diameter or larger were used for the tests just described. Other tests with 0.33 mm diameter wire in argon and vacuum showed increased drift, with several failures prior to 10,000 hours of exposure. It was estimated that for vacuum and argon drift can be expected to reach -65°C per 10,000 hours for 0.33 mm diameter wire, compared with -23°C per 10,000 hours for the 0.51 mm size. Also, failure may occur in 2000 to 4000 hours for both sizes in vacuum; in argon erratic behavior of the smaller diameter wire can be expected after 4000 hours. These estimates were all for an exposure temperature of 1327°C.

Darling, et al. (1970, 1971) discussed the reactivity of platinum and platinum-rhodium alloys with powdered refractory oxides under conditions of unusually low oxidizing potential. The work is described here under Pt-10Rh vs Pt. These investigators recommended the use of magnesium oxide because of its inertness compared with Al_2O_3, ThO_2, and ZrO_2 when the oxygen level was very low, such as the case with gettered argon or high vacuum. Magnesium oxide also aided thermocouple performance in reducing atmospheres. A thermocouple held at 1450°C for 450 hours, while in an atmosphere of dissociated ammonia and packed in fine magnesium oxide powder, drifted in the negative direction by less than 10°C apparent change in temperature. A similar thermocouple insulated with aluminum oxide showed a much greater drift.

Work concerning the effect of stress on thermoelectric power was reviewed by Morgan (1968), who also described his own analytical and experimental investigations. The analytical results enabled calculation of the voltage hysteresis loop observed with metal sheathed insulated couples, apparently caused by thermal stresses during temperature cycling. A calculated and an experimental thermal cycle between 20 and 550°C gave maximum output voltage discrepancies of 23 μV and 14 μV, respectively.

The possible effects of stress and other contributing agents to deterioration were considered by Hancock (1967) for a nuclear application. Sheathed couples were used, with magnesium oxide insulation, and with ungrounded junctions. The sheath consisted of a 15 to 23 cm length of Pt-13Rh at the hot junction end, followed by an additional 3 meter length of stainless steel. The couples operated in a neon atmosphere with graphite present, and with a thermal neutron flux of about 10^{14} n/cm^2 sec. Satisfactory operation for 40 to 180 days at a temperature of about 1300°C was reported. Open circuit failures occurred after this time. These were believed to be caused by a combination of grain growth embrittlement of the platinum conductor and stresses introduced by the differential expansion of the stainless steel and the wires. Errors during operation of the couples caused by transmutation of the thermocouple elements were estimated at not more than 5 per cent per 10^{20} n/cm^2 integrated thermal flux.

Dau and his associates (1968) made a brief review of papers reporting on short-term changes in calibration that are proportional to nuclear heating. Early work by Madsen (1951) was quoted regarding such effects on Pt-13Rh vs Pt couples in ceramic sheaths and in a thermal neutron flux of 10^{12} n/cm^2 sec. Dau, et al., concluded from this and reports of work with other thermocouple combinations that, generally, dose rate effects on calibration are less than 1 per cent for flux under 10^{13} n/cm^2 sec.

Pt-13Rh vs Pt-0.5Rh.   This pair has been mentioned by Bennett (1961, p. 27) and Billing (1963) as a replacement for Pt-13Rh vs Pt, because of the improvement in drift characteristics obtained by substituting the 0.5 per cent rhodium alloy for platinum. More extensive information is available on a couple using a 1 per cent rhodium alloy, discussed below.

Pt-13Rh vs Pt-1Rh.   The curve in Figure 2-3 was plotted from data given by Zysk (1962) on the voltage of each alloy vs National Bureau of Standards Platinum 27. The voltage output is less than that of Pt-13Rh vs Pt. From these same data the thermoelectric power in the 700-1700°C region is about 10 to 12 $\mu V/°C$.

Metcalf (1950) substituted Pt-1Rh for pure platinum after experiencing drift with Pt-13Rh vs Pt. The drift was believed to have been caused by transfer of rhodium from the alloy into the platinum conductor via the vapor phase. It was pointed out that the presence of only 0.004 per cent rhodium in the platinum causes the output of the Pt-13Rh vs Pt couple at 1200°C to decrease by 70 μV or more than –0.5 per cent. That same addition of rhodium to a platinum-rhodium alloy represents a relatively small increment of rhodium composition and the voltage of the thermocouple is not as greatly altered by the change. Such a thermocouple is relatively insensitive to rhodium migration either via the vapor phase process or from diffusion at the junction.

This type of thermocouple was used in vacuum after fitting a curve to three calibration points obtained at the melting points of silver, cobalt, and palladium. After 4 hours at 1510°C, the output of the Pt-13Rh vs Pt-1Rh couple drifted by –0.8°C; after 2 hours a Pt-13Rh vs Pt couple changed by –6°C. The couples were enclosed in aluminum oxide insulators.

Metcalf found no direct evidence of additional rhodium[6] in the negative wires of used Pt-13Rh vs Pt thermocouples, although he reported increased trace amounts of copper, silicon, and magnesium, apparently in both wires. Later work by Walker, Ewing, and Miller (1962; 1965, p. 601) suggested the possibility of iron contamination of the platinum from impurities in the insulators, but Metcalf did not detect an increase in the amount of that element during spectrographic tests. Regardless of the contamination mechanism it is evident that considerable improvement resulted from the use of Pt-1Rh rather than platinum.

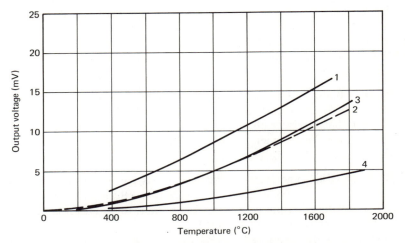

FIGURE 2-3 All-alloy platinum-rhodium thermocouples. 1. Pt-13Rh vs Pt-1Rh, Zysk (1962). 2. Pt-20Rh vs Pt-5Rh, Bedford (1964). 3. Pt-30Rh vs Pt-6Rh, Burns and Gallagher (1966). 4. Pt-40Rh vs Pt-20Rh, Bedford (1965).

Hancock (1967) substituted this combination for Pt-13Rh vs Pt and made other improvements after open circuit failures occurred with the latter in metal sheathed oxide insulated cables. It was believed that the use of the Pt-1Rh alloy resulted in less grain growth, with its subsequent tendency for failure, than was the case with pure platinum.

Pt-18Rh vs Pt-13Rh. A voltage-temperature curve was included in a patent by Nishimura (1958). The data were used for comparative purposes with other combinations.

Pt-20Rh vs Pt-5Rh. According to Bennett (1961, p. 26) this thermocouple was used by D. Hanson in 1927 for melting point determinations. In England the couple is sometimes referred to by that name. Apparently there was little further work done until Metcalf's results with Pt-13Rh vs Pt-1Rh, which encouraged Welch (1956) to adopt Pt-20Rh vs Pt-5Rh for use in phase equilibrium studies by means of thermal analysis. The hot junction was directly exposed to various mixtures of oxides in the $CaO-MgO-Al_2O_3-SiO_2$ quaternary system at temperatures in the 1300-1700°C range, presumably in air. The near linearity of the voltage-temperature curve above 1400°C was reported to be an advantage for calibration work, and the advantage of room temperature cold junction insensitivity was pointed out. Other applications were described by Chaston (1957) and by Sharp (1963). The latter work concerned temperature measurements in molten steel during continuous immersion times of 20 minutes at 1700°C.

Bedford (1964) conducted very detailed calibration work with 13 couples from four lots of wire and produced the reference tables now used for the 0-1800°C range. Data from the tables are plotted in curve 2 of Figure 2-3. Using the tables, the thermoelectric power is 8.5 μV/°C at 1000°C, 10 μV/°C at 1400°C, and 9.8 μV/°C at 1700°C.

Bedford discussed the accuracies attainable by the user after making a calibration and employing the difference curve method. The resulting calibration was estimated to be accurate to ±1°C below 1063°C, ±2°C from 1063 to 1552°C, and ±3°C above. The figures assume the use of the six freezing points of zinc, antimony, silver, gold, palladium, and platinum, for establishing first a calibration curve, and then the difference curve, using the tables. It was estimated that there would be little loss in accuracy by omitting two of the three fixed points below the gold point. Bedford concluded that this type of couple is as accurate as Pt-10Rh vs Pt above 1000°C and can be used over an even wider range in the high temperature region.

The thermoelectric power is low in the room temperature region. It was pointed out that this makes cold junction compensation extremely easy. A nominal correction within ±3°C based on the reference tables would ensure that the error above 1000°C was not greater than 0.1°C. The error would not exceed 1°C even if no correction were made for a reference junction at room temperature. This same lack of sensitivity at lower temperatures is also a disadvantage when it is necessary to measure temperatures in this region. The lower temperature limit cannot be said to match that of Pt-10Rh vs Pt because of this condition.

There was general agreement by Chaston (1957), Bedford (1964), and Zysk (1964) that a maximum continuous operating temperature of 1700°C in air is reasonable. Kostkowski (1960), and Kostkowski and Burns (1963) estimated that error after 12 hours at that temperature would not be greater than 4°C. The ultimate limit for short term use is set by the Pt-5Rh alloy melting point which appears to be slightly lower than 1800°C, according to Bedford. Generally, performance in various atmospheres or in vacuum can be expected to at least equal that of Pt-10Rh vs Pt.

Bedford also discussed stability observed during his calibration work. Repeated measurements of the gold point showed no instability. Small instabilities were present during calibration by comparison with a couple of Pt-10Rh vs Pt. The maximum spread in the data for as many as 14 comparisons of any one couple with Pt-10Rh vs Pt in the 0-1500°C range was about 1°C at each temperature. It was suggested that at least some of the spread was due to the instability of the Pt-10Rh vs Pt couple rather than the Pt-20Rh vs Pt-5Rh pairs.

Drift tests were also performed by maintaining aluminum oxide sheathed couples at 1700°C in air for 200 hours. Negative drifts of up to -5°C were observed by the end of this period. The results are comparable with drift data by Walker, Ewing, and Miller (1962) for Pt-10Rh vs Pt.

Pt-30Rh vs Pt-1Rh.   This pair was used by Thomas (1963) for calibrating refractory metal couples in the 100-1000°C interval, after first calibrating by comparison with a standard Pt-10Rh vs Pt thermocouple. The uncertainty of the comparison calibration was estimated at ±0.5°C.

Pt-30Rh vs Pt-6Rh.   A thermocouple with this composition was first developed for commercial applications by a German firm, and according to Rudnitskii and Tyurin (1956) it was available in 1953. The early calibration tables for this couple were revised by Obrowski and Prinz (1962). These tables have been referred to as the old and the new Degussa reference calibrations, respectively, after the German manufacturer. Calibration data summaries were given by Caldwell (1962) and Zysk (1964) in review articles.

A reference table was later developed by Burns and Gallagher (1966), based on work with 11 thermocouples from four different manufacturers, and covering a span of 0-1820°C, the latter temperature being approximately the liquidus value for the Pt-6Rh alloy. This couple can be used for several hundred hours at 1750°C, and intermittent use to 1800°C is considered possible.

Improved stability, increased mechanical strength, and higher maximum operating temperature were cited by Burns and Gallagher as factors which make Pt-30Rh vs Pt-6Rh superior to Pt-10Rh vs Pt above 1000-1100°C, where the thermoelectric power is comparable with that of the latter couple. Further, this combination was said to offer the most favorable over-all characteristics of the "family" of all-alloy couples which also includes Pt-20Rh vs Pt-5Rh and Pt-40Rh vs Pt-20Rh. Zysk and Robertson (1971) indicated that in the United States this couple has been chosen as the standard all-alloy pair for temperatures above 1500-1600°C up to the 1760°C region. Differences among these all-alloy pairs are small, nevertheless, and the Pt-30Rh vs Pt-6Rh couple has just slightly higher thermoelectric power at high temperature, and somewhat higher tensile strength, than Pt-20Rh vs Pt-5Rh. A curve of voltage versus temperature is included in Figure 2-3. The thermoelectric power S is 7.7 μV/°C at 800°C, 10.4 μV/°C at 1200°C, and 11.7 μV/°C at 1600°C. The magnitude of S goes through a maximum at about 1600°C and is very near the value for Pt-10Rh vs Pt at that point.

Regarding the use of the reference tables, Burns and Gallagher indicated that calibrations of a particular Pt-30Rh vs Pt-6Rh thermocouple at four points, about 600, 1063, 1300, and 1552°C, are sufficient for construction of an accurate curve. The first three points are determined by comparison with a standard Pt-10Rh vs Pt couple, the fourth by the melting wire method at the palladium point. The resulting curve can then be used with the reference tables to construct a difference curve; the resulting calibration will then be accurate to within ±6 μV up to 1063°C, the equivalent of about ±3°C up to 1550°C, and of about ±5°C at higher temperatures. Calibration above 1600°C for most applications work may be undesirable because some drift would be introduced which would render the couple less reliable for use. Calibration values above 1600°C can be obtained accurately by extrapolating the difference curve. Generally, manufacturer-supplied thermocouples are within ±0.5 per cent of agreement with the tables in the 500-1800°C range, and within ±15 μV at lower temperatures.

According to Starr and Wang (1969) reference grade wires are available with a ±0.25 per cent tolerance in the range of 870-1650°C. In the United States the designation for commercially supplied wires is Type B for this combination. For applications where the cold junction temperature is above about 50°C, extension wires are available which consist of a copper and a copper-nickel alloy pair. Copper wires, giving no compensation, are satisfactory below 50°C. In most cases the cold junction temperature need not be determined, since the thermoelectric power and voltage are very low in the room temperature region up to about 50°C.

Gaylord and Compton (1968) tested Pt-30Rh vs Pt-6Rh thermocouples for 100 hours in slowly moving air at 1320°C, finding a maximum drift of about 8°C from the initial indicated temperature. There were fluctuations in the magnitude of the drift, with maxima and minima. Next, two couples were exposed for 28 hours to the hot combustion products of a turbine atmosphere simulator. The simulator burned JP-4 fuel, the temperature varied from 1010 to 1260°C, and gas velocity was approximately 210 meters per second. The maximum drift, which occurred for one thermocouple after mechanical breakage and repair, was about -17°C at 1316°C. The other unit showed a drift of -3°C at that temperature.

Earlier work by Rudnitskii and Tyurin (1956) seemed to indicate that bare wire couples were unstable at 1800°C, but the more encouraging results of other investigators suggest that some unusual condition existed in the former instance. Rudnitskii and Tyurin also investigated aluminum oxide sheathed wires of a number of thermocouple alloys, including Pt-30Rh and Pt-6Rh, at lower temperatures. Drift rates were generally higher than would be expected using more recent results; in this case, there was the possibility of iron impurities existing in the insulators, and subsequent contamination-produced drift in the output may have occurred. This work was done prior to the discovery of that effect by Walker, Ewing, and Miller (1962; 1965, p. 601).

Walker, et al., made an investigation of thermocouple drift, which led to the conclusion that this couple is less susceptible to iron contamination than Pt-13Rh vs Pt-1Rh or Pt-10Rh vs Pt. It was estimated that after 240 hours in air at 1384°C, thermocouples enclosed in low iron content aluminum oxide insulators would change in calibration by less than 0.2°C as compared with –1.4°C for Pt-10Rh vs Pt under similar conditions. A 120-hour exposure in either air, argon, or vacuum, with the best quality aluminum oxide insulators, resulted in no change in calibration as measured at 860°C; the test exposure temperatures were 1000 and 1200°C. This was also true for air exposure at 1600°C. However, in argon and vacuum at this temperature the changes were –7 and –2°C, respectively, after 120 hours and when measured at 860°C. Drift was affected by wire size for atmospheres of argon and vacuum; 0.51 mm diameter wires were tested.

Hendricks and McElroy (1966, 1967) reported on work with thermocouples insulated by two-hole aluminum oxide sheaths which had been treated to remove as much of the iron impurities as possible. These units were exposed to a temperature of 1300°C in a vacuum of 2×10^{-7} torr for 352 hours. Drift was within ±5°C. After an additional 1012 hours at 1450°C, drift was within ±8°C; other tests confirmed these results. The amount of drift was not affected by wire size. One test at 1450°C demonstrated that drift occurs when the hot junction is in contact with niobium and a large temperature gradient along the wires exists. The drift, which amounted to about –80°C during 1012 hours at 1450°C, was attributed to diffusion of niobium into the hot junction.

Glawe (1970), and Glawe and Szaniszlo (1971) described long term drift tests of Pt-30Rh vs Pt-6Rh thermocouples with 0.51 mm wire diameters in argon and vacuum. The wires were enclosed in aluminum oxide insulators of at least 99.5 per cent purity. These thermocouples completed a 10,000-hour test in argon at 1327°C, and a negative drift of the indicated temperature was observed which ranged from –7.2 to –18.1°C. The average drift was –13°C, compared with –22°C for Pt-13Rh vs Pt couples tested under the same conditions. Drift in vacuum was also observed, and there were two failures prior to 3000 hours of exposure. It was estimated that drift in vacuum of –26°C per 10,000 hours with failure after 2000 to 4000 hours can be expected. In this case, the drift and expected time to failure were about the same as for Pt-13Rh vs Pt.

Selman (1971), and Selman and Rushforth (1971) found that properly prepared Pt-5Rh sheathed thermocouples of this type showed less drift than comparable Pt-13Rh vs Pt couples, also sheathed. Couples in magnesium oxide insulation with hermetically sealed sheaths, and with an internal atmosphere of argon, can be expected to drift about –2 to –5°C after 500 hours at 1450°C.

Couples were subjected by Briggs and Johnson (1966) to 14 temperature cycling tests ranging from 25 to 1650°C in an atmosphere of helium. The total test time was about 425 hours. These couples, which were insulated with crushed beryllium oxide and sheathed in tantalum, showed no appreciable drift. Some failures were observed, apparently caused by thermal stresses.

Quick immersion thermocouple assemblies have been developed for the steel industry, and these units were reported by Ramachandran (1963) and Ramachandran and Acre (1964) to be useful up to the melting point of the Pt-6Rh alloy. Steel industry applications in Russia have been mentioned by Fedorovskii (1969). Kislyi and coworkers (1968) discussed the temperature measurement of flue gases, molten copper, and slag, where a silicon carbide sheath was used for protection of an aluminum oxide inner sheath, which in turn encased the thermocouple. This arrangement was said to protect the couple for at least 100 hours at 1400°C during furnace tests. Without the inner sheath there was no thermocouple drift exceeding ±25 μV for 100 hours at either 900 or 1100°C; failure by embrittlement resulted after five hours at 1200°C, apparently due to contamination by silicon vapor which originated from the silicon carbide.

Pt-40Rh vs Pt-3Rh.   Zysk, et al. (1969) described thermocouples of this nominal composition in a patent. The negative conductor was either the binary alloy of platinum and rhodium or these elements plus from zero to 2 per cent of a material selected from the following group: zirconium, titanium, gold, cerium, thorium, and their oxides. A maximum operating temperature of about 1750°C was given. Extension wires which matched this couple over the 0 to 1000°C range were also described.

Pt-40Rh vs Pt-20Rh.   Jewell, Knowles, and Land (1955) were first to report on applications of this couple. It was developed for use in liquid steel temperature measurements above the range of Pt-13Rh vs Pt. Because of sheath compatibility problems for immersion in liquid metals at nearly 1900°C the arrangement did not prove to be fully satisfactory. The thermocouple, nevertheless, showed promise for other uses. Its principal advantage over other platinum-rhodium thermocouples was the exceptionally high temperature range of up to 1850°C, and even to 1880°C in special cases.

Later calibration work was reported by Zysk (1960, 1962, 1964), St. Pierre (1960), and Yust and Hall (1966). Bedford (1965) undertook preparation of the reference tables which are now in use. These tables are based on the calibration of couples assembled from four lots of wire from three suppliers. A total of ten thermocouples were calibrated in air at the freezing points of zinc, antimony, silver, and gold. A calibration point at the melting point of palladium was also taken, using an argon atmosphere. Palladium and platinum melting point calibrations were later made after comparison calibration work using Pt-10Rh vs Pt and Pt-20Rh vs Pt-5Rh thermocouples. The completed tables span the 0-1880°C interval; the melting temperature of the lower rhodium content alloy is near 1880°C. It was concluded that if a thermocouple calibration is based on a difference curve with respect to the tables and to a curve made through six well distributed calibration points, then the resulting calibration accuracy will be ±1°C below 1063°C, ±2°C from 1063 to 1552°C, and ±3 to 4°C above 1552°C.

The thermoelectric power S was found to be constant within ±1.5 per cent between 1550 and 1880°C. The resulting near linearity is an advantage pointed out earlier by St. Pierre (1960). Representative values of S, from a comprehensive table of S versus temperature by Bedford (1965), are 2.24 $\mu V/°C$ at 800°C, 3.55 $\mu V/°C$ at 1200°C, 4.48 $\mu V/°C$ at 1600°C, and 4.53 $\mu V/°C$ at 1800°C. The value of S goes through a maximum at 1725°C. Voltage data from the tables are plotted in Figure 2-3.

Bedford suggested use of the Pt-40Rh vs Pt-20Rh couple for temperatures above 1500°C in general, or above 1000°C for special cases. This rather unusual low temperature limitation is due to the low thermoelectric power below 1500°C. One advantage of a low value of S is a lack of sensitivity to cold junction temperature errors, although this condition is not as pronounced as for the Pt-20Rh vs Pt-5Rh couple. In order to prevent a measurement error greater than 0.1°C at 1000 and 1800°C, a room temperature cold junction must be known to within 0.5 and 1°C, respectively. The sensitivity to cold junction error was said to be about five times that of the Pt-20Rh vs Pt-5Rh thermocouple.

The high temperature stability was considered good, as evidenced by the agreement of repeated measurements during the calibration process. Drift was evaluated by 480 hours of air exposure at 1700°C and by measuring the output at 1552°C. Three couples were tested; all were enclosed in aluminum oxide insulators. The drift results were considered unusual, because there were indications of a positive drift that reached a maximum value at about 200 hours, followed by a negative direction drift which subsequently resulted in negative deviations from the original calibrations. The temperature drifts were rather small, being within ±4°C.

Zysk (1964) quoted drift results from unpublished work by Toenshoff and Wisely, and at high temperatures the agreement with Bedford was good. After four couples were exposed at 1700 or 1800°C for periods of time ranging from 170 to 772 hours, the maximum observed drift measured at 1500°C was about -4°C; measured at 1600°C the drift was about -6°C. The drift at lower measurement temperatures was higher than these figures, however, the maximum value measured at 1000°C being about -16°C. In these experiments the wires were sheathed with aluminum oxide insulators up to a point about a centimeter from the hot junction, leaving the remaining length of wire and the junction exposed to still air.

Rh vs Pt.   Calibration data for rhodium relative to National Bureau of Standards Pt 27 are plotted from information by Zysk (1962) in Figure 2-4. A comparison with Pt-10Rh vs Pt may be made with the aid of Figure 2-2. The thermoelectric power, from curves of absolute thermoelectric powers of rhodium and platinum by Vedernikov (1969), is about 22 $\mu V/°C$ at 1100°C and 25 $\mu V/°C$ at 1400°C.

Rudnitskii and Tyurin (1956) conducted drift tests on rhodium as well as on a number of platinum-rhodium alloys. The resulting data indicated that rhodium was less susceptible to drift in air than the alloys. The wires were within aluminum oxide insulators, and the possibility exists that iron impurities were present and were a factor in determining drift. Walker, Ewing, and Miller (1962; 1965, p. 601) later demonstrated that drift is often the result of iron impurity contamination, and showed that rhodium is less affected than platinum.

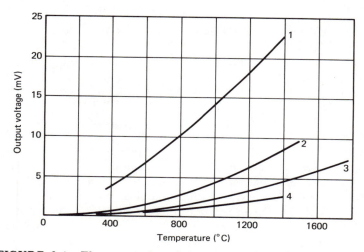

FIGURE 2-4   Thermocouples of rhodium with platinum and platinum-rhodium alloys. 1. Rh vs Pt, Zysk (1962). 2. Rh vs Pt-10Rh, Schulze (1939). 3. Rh vs Pt-20Rh, Rudnitskii and Tyurin (1960). 4. Rh vs Pt-30Rh, Zysk (1962)

Walker, et al., exposed a 0.51 mm wire diameter aluminum oxide sheathed thermocouple of Rh vs Pt for 240 hours in air at 1384°C; a calibration change of −1.2°C was determined at that temperature. Additional experiments in air, argon, and vacuum using low iron content aluminum oxide insulators[7] showed no change in calibration after 120 hours at 1000 or 1200°C, when voltage was measured at 860°C. In other tests, drift of +1 and −5°C were observed after 120 hours in vacuum at 1380 and 1600°C, respectively. Again the drift measurement temperature was 860°C. No change was observed for air or argon atmospheres at 1380°C. The results were similar to those for Pt-10Rh vs Pt, since platinum was used in both thermocouples and had been shown to be the most susceptible to contamination of the metals in the platinum-rhodium system.

Rh vs Pt-10Rh.   Schulze (1939) gave a calibration table, including thermoelectric power, and attributed the work to Henning (1938). Data from the table are plotted in Figure 2-4 and can be compared with Rh vs Pt. The thermoelectric power was given as 8.4 μV/°C in the 900-1000°C region and 11.4 μV/°C in the 1400-1500°C interval. No information was given for higher temperatures. It was shown in the table that the thermocouple voltage is only 1 μV at 100°C. Therefore no cold junction reference is needed when high temperatures are measured.

Rh vs Pt-20Rh.   This thermocouple was investigated by Rudnitskii and Tyurin (1956, 1960); some of their calibration data are given in Figure 2-4. The thermoelectric power S is low at room temperature, and it was pointed out that for uncorrected cold junction temperatures below 100°C, measurement error at 1800°C would be less than 1 per cent. The value of S is on the order of 4 μV/°C near 900°C, 6 μV/°C at 1300°C, and 8 μV/°C at 1800°C.

This thermocouple appeared to be less susceptible to drift than Pt-30Rh vs Pt-6Rh during early development work by Rudnitskii and Tyurin, although the output and thermoelectric power were rather low. During later tests the thermocouples were first annealed for two hours at 1200°C in a platinum wound furnace, and then mounted in aluminum oxide insulators for experiments at 1550°C in air. A 1000-hour drift test at that temperature resulted in an initial downward drift of -2.9°C after 100 hours, followed by a near return to zero drift during a period of 500 additional hours. Output again decreased after 600 hours of testing and the net negative drift after 1000 hours was -12°C.

Other couples were tested at 1800°C in three types of insulating materials, and the drifts were approximately 0.5 per cent (about 9°C) or less during 100 hours. The results were characterized by initial output drift, which reached a maximum magnitude, followed by an approach to the original output value. The maximum drift in aluminum oxide with 1 per cent titanium dioxide was +0.53 per cent at 50 hours and +0.18 per cent at 100 hours. With Sinoxal, an aluminum oxide containing additives of silica and calcium oxide, the drift was -0.38 per cent at 50 hours and -0.26 per cent at 100 hours. With beryllium oxide, drift was -0.32 per cent at 7 hours and zero at 100 hours.

Aluminum oxide sheathed thermocouples were exposed to air for 240 hours at 1384°C by Walker, et al. (1962) with a change in calibration of -3.5°C resulting at that temperature. The drift occurred during the first few hours of the test, reached -3.5°C after 120 hours, and then remained constant.

Rh vs Pt-30Rh.   Rudnitskii and Tyurin (1956) evaluated this pair during their work with other thermocouples, and found the output to be even lower than that of Rh vs Pt-20Rh. The output was said to be 4.6 mV at 1800°C. This was compared with 6.75 mV at 1800°C for Rh vs Pt-20Rh. Tabulated calibration data gave slightly different but comparable figures for 1800°C. A voltage-temperature curve is shown in Figure 2-4, which was constructed from separate data by Zysk (1962) for the two metals relative to platinum. The thermoelectric power is in the order of 3 μV/°C in the 1200° region.

This combination was found to be more stable than that of Rh vs Pt-20Rh. Under the same conditions of bare wire exposure to air at 1800°C, the Rh vs Pt-30Rh couple showed a -3.7 per cent output change after 2 hours, compared with -5 per cent for Rh vs Pt-20Rh. Since a Pt-30Rh vs Pt-6Rh couple which was also tested showed excessive drift under the same conditions and is now known to normally perform well, these figures should be used only comparatively, owing to what was apparently an unusual contamination condition. This was not the more typical problem of iron contamination from thermocouple sheaths since the wires were bare.

Ruthenium and Platinum

General Information. The melting point of ruthenium is 2250°C, well above platinum at 1772°C and rhodium at 1963°C. The oxidation and volatilization processes associated with platinum and rhodium at high temperatures in air are also characteristic of ruthenium and ruthenium oxide. However, the rate at which these processes take place is much greater for ruthenium. According to Israël (1966), the oxidation of ruthenium from Pt-Ru alloys occurs preferentially along the grain boundaries. This causes brittleness and makes these alloys unsuitable for use in high temperature oxidizing atmospheres. A 10 per cent weight loss of the ruthenium present in Pt-10Ru takes place during approximately 24 hours exposure to air at 1000°C.

The inferior oxidation properties of ruthenium alloys discouraged their use, until nuclear application requirements prompted a search for radiation-insensitive thermocouples. Ehringer and his associates (1966) computed the effects of radiation on platinum-ruthenium alloys and concluded that such alloys, together with platinum, can be used in an integrated neutron flux of up to 10^{21} n/cm^2 without transmutation effects. This result is based on the low thermal neutron cross section of ruthenium, which is less than 2 per cent of that of rhodium.

Curves of thermoelectric voltage of platinum-ruthenium alloys vs platinum, plotted against ruthenium alloy content at a specified constant temperature, generally show a rapid voltage rise for ruthenium additions up to some 5 to 9 per cent. Curves from Schulze (1939), Stauss (1941), and Vines and Wise (1941) show this characteristic. Israël (1966) investigated the 0.5 to 10 per cent ruthenium range and found that except at temperatures of 1500°C and above, plots of voltage versus alloy content show broad maxima at less than 10 per cent ruthenium content. At 900°C the maximum is at about 6 per cent ruthenium; at 1200°C the maximum is at a little less than 8 per cent ruthenium. The thermoelectric power S of the dilute alloys is somewhat less influenced by temperature than for most metals. For Pt-1Ru, S is nearly constant at about 7.6 μV/°C over the entire 0-1600°C temperature range.

Pt-6Ru vs Pt. Israël (1966) advised the choice of the Pt-6Ru alloy for use with platinum as a thermocouple for nuclear environments. The selection of that particular alloy was based on the increasing difficulty of working Pt-Ru alloys when the ruthenium content is further increased, plus the fact that in the 6 per cent range minor variations of alloy content do not result in serious calibration differences. There is also no appreciable gain in thermoelectric power by further increasing the alloy content, since the maximum value of S occurs in the 6 to 8 per cent region, depending on the temperature.

Calibration work was done in air to $1000°C$ and in nitrogen to $1600°C$. Figure 2-5 is plotted from the calibration tables. The voltage-temperature curve for Pt-10Rh vs Pt is included for comparison. The thermoelectric power of Pt-6Ru vs Pt is about $20~\mu V/°C$ at $1000°C$. The pronounced weight loss of ruthenium from the platinum-ruthenium alloy required the use of the nitrogen atmosphere above $1000°C$.

MULTICOMPONENT SYSTEMS INCLUDING PLATINUM

General Information

Several of the thermocouples in this section utilize alloys of gold and palladium which are negative in thermoelectric output relative to platinum. The earliest investigation of the gold-palladium alloy system is generally attributed to Geibel (1910), who found that the greatest negative output was reached with an alloy of Au-40 Pd. The voltage at $1000°C$ relative to platinum was –42.7 mV, a large output for a noble metal thermocouple.

The property of hydrogen absorption by palladium is well known. Schulze (1939) reviewed the phenomenon and stated that the thermoelectric voltage is affected by the absorbed gas. Nevertheless, palladium can be used in thermocouples in other atmospheres. Also important is the fact that the alloying of palladium with other metals greatly reduces the absorption capacity, so that the Au-40Pd alloy is little affected. A ternary alloy of palladium with platinum and gold has been used in a hydrogen atmosphere for extended periods with little thermoelectric change.

Thermocouples with Three or More Elements

Pt-10Ir vs Au-40Pd. This combination was first investigated by Geibel (1910) and was used for many years, but fell into disfavor when excessive drift characteristics at high temperature became known. Bennett (1960) described later work where it was found that such characteristics can be eliminated by proper treatment of the Au-40Pd wire. This is done by preventing appreciable gas absorption during alloy preparation and by use of a stabilizing treatment prior to drawing.

Accinno and Schneider (1962) described the drift characteristic as the result of an improper pre-annealing treatment of the alloy.

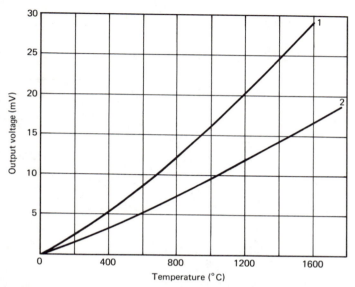

FIGURE 2-5   Platinum-ruthenium and platinum-rhodium thermo-couples.   1. Pt-6Ru vs Pt, Israël (1966).   2. Pt-10Rh vs Pt, Shenker, et al. (1955).

The properly prepared thermocouple wires are available commercially under the tradename of Pallador I.[8] The voltage-temperature curve is within ±1 per cent of the curve for iron-constantan over the 0 to 1000°C range recommended by the manufacturer. An accuracy of ±2°C can be obtained over the full range using the curve supplied with each wire shipment. A curve from other data by Zysk (1964) is shown in Figure 2-6. The output is linear above 400°C and the thermoelectric power is approximately 63 μV/°C.

According to Bennett (1960), Pallador I can be used in air to 1000°C for prolonged service without embrittlement or significant change in calibration. The solidus points of the alloys are considerably higher than this figure. A value of 1450°C was given for Au-40Pd. For Pt-10Ir the solidus point is 1780°C. In other work, Bennett (1961, p. 31) stated that this couple is widely used as the functional element in high frequency electric meters, and is also useful in biology where resistance to tarnishing is important. Powell, Caywood, and Bunch (1962), and Powell and Caywood (1962) reported that the voltage-temperature characteristics in the cryogenic region are similar to those of Chromel vs Alumel, but gave no quantitative data. Inhomogeneities observed during cryogenic tests were said to be larger than those of Chromel-Alumel, so no complete calibration was made.

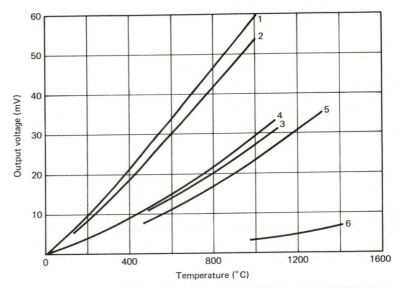

FIGURE 2-6   Multicomponent thermocouples containing platinum, I.   1. Pt-10Ir vs Au-40Pd, Zysk (1964).   2. Pt-10Rh vs Au-40Pd, Zysk (1964).   3. Pt-10Ir vs Pd, Zysk (1962).   4. Pt-15Ir vs Pd, Ihnat (1959).   5. Pt-20Rh vs Pd, Zysk (1962).   6. Ir vs Pt-10Rh, Zysk (1962).

Pt-10Rh vs Au-40Pd.   Zysk (1964) described this pair as an older combination, still in use, and gave information which is plotted in Figure 2-6. The output is only slightly less than that of Pt-10Ir vs Au-40Pd and is linear above 500°C, with a thermoelectric power of 58 $\mu V/°C$ in the 500-1000°C range. According to information quoted by Sagoschen (1961) the couple may be used up to 1000°C.

There is no specific reference to the drift problem experienced with the earlier version of the Au-40Pd alloy, discussed under Pt-10Ir vs Au-40Pd. Presumably, use of the properly prepared alloy would avoid the condition.

West (1960) chose this couple for an application in calorimetry where high sensitivity is desirable to measure small temperature differences, and where it was also desired to avoid assembly problems associated with base metal couples. The installation consisted of a 20 junction thermopile with gold extension wires. The Au-40Pd wire was reported to be stiffer and more difficult to handle than normally used thermocouple wires, but was more easily soldered or welded.

Pt-2Ru vs Pt-8Au.   These alloys were selected by Hill (1957) for use in a transient method of measuring thermal conductivity. The thermoelectric power was about 15 μV/°C.

Pt-10Ir vs Pd.   Ihnat (1959) investigated this thermocouple for use in jet engine temperature measurements, but later chose Pt-15Ir vs Pd for the final selection. The voltage-temperature curve is plotted in Figure 2-6 using tabulated data by Zysk (1962). The curve is very similar to that of Pt-15Ir vs Pd, and the thermoelectric power is 35 μV/°C in the region of 1000°C.

Clark and Hagel (1960) and Ihnat and Hagel (1960) described tests of a thermocouple which was exposed in a combustion chamber using fuel oil containing 0.4 per cent sulphur, for simulation of jet engine combustion atmospheres. During 600 hours total exposure time at 1260°C, the combined instability and drift did not exceed ±0.5 per cent of the initial output. The output deviation was an irregular function of time rather than a smooth drift, and drift appeared somewhat more pronounced than for Pt-15Ir vs Pd.

The couple appears suitable for use in oxidizing atmospheres. Ihnat did not recommend its use in hydrogen. Palladium is known to absorb hydrogen and a corresponding change of thermoelectric power takes place.

Pt-15Ir vs Pd.   Ihnat (1959) was the first to study this thermocouple and give a calibration table. A curve plotted from data in Ihnat's report is given in Figure 2-6. Freeze, Caldwell, and Davis (1962) and Freeze (1963) later described calibrations of six thermocouples which were used in preparing reference tables for the –62 to +1400°C range. The maximum spread of voltage of the six couples from the averaged values of the tables was 0.4°C at 260°C and 1.5°C at 1371°C. Tabulated figures on thermoelectric power were also given which showed an increase from 16.6 μV/°C at 0°C to 43 μV/°C at 1200°C.

Caldwell (1962) suggested that the upper useful limit of the couple would be somewhat below the 1554°C melting point of the palladium. At lower temperatures palladium is superficially oxidized when heated to 700°C in air. Above 875°C the palladium oxide decomposes leaving a bright metal surface. A gray to black film forms on the Pt-15Ir at about 900°C and disappears at a temperature 200 to 300°C higher.

Wire sizes down to 0.25 mm have been used and wires as small as 0.025 mm could probably be prepared, because of the ductility of both metals. Olsen (1963) described an annealing schedule which consisted of heating the alloy wire for one minute at 1300°C; the palladium wire was held for 25 minutes at 750°C, after momentarily heating to about 1000°C to remove a dark oxide coating.

Wire breakage at the hot junctions was observed by Clark and Hagel (1960), who attributed the problem to the fact that the temperature coefficient of expansion of palladium is about twice that of the Pt-15Ir. This results in thermal stresses during heating. It was necessary to design a stress-relieving loop in the wires to accommodate the expansion problem. Each application of this couple must be considered with thermal stress limitations in mind in order to eliminate the condition.

A test in which the combustion atmosphere in a jet engine was simulated was described in reports by Ihnat (1959), Clark and Hagel (1960), and Ihnat and Hagel (1960). Fuel oil containing 0.4 per cent sulphur was burned and drift from the initial calibration was determined by comparison with a Pt-10Rh vs Pt thermocouple at intervals of 25 or 50 hours in the combustion atmosphere. The drift did not exceed ±0.35 per cent during a 400-hour exposure at 1260°C. The drift appeared as a fluctuating deviation, generally in the positive direction, rather than as a steadily increasing error. Drift tests in air were also described by Clark and Hagel. A 1000-hour exposure test to air flowing at Mach 0.1 and at a temperature of 1093°C[9] was conducted. Drifts of ±0.2 per cent were observed. Clark (1962) later compared simulated jet engine test results with those for a Chromel-Alumel type couple and indicated that the noble metal pair showed superior performance at 980 and 1200°C.

Pt-20Rh vs Pd.   Voltage-temperature data for each metal versus Platinum 27 were given by Zysk (1962), from which the curve in Figure 2-6 was constructed. The thermoelectric power, determined from the slope of the curve, is about 26 μV/°C at 600°C and 38 μV/°C at 1200°C.

An investigation of this pair was reported by Ihnat (1959) and Ihnat and Hagel (1960) during the work previously described under Pt-15Ir vs Pd. There was somewhat less drift than with Pt-15Ir vs Pd during tests in a simulated jet engine combustion atmosphere. During a 600-hour test at 1316°C, output deviations remained within about ±0.15 per cent.

Ir vs Pt-10Rh.   Tables from Zysk (1962) showed voltage versus temperature for each metal relative to Platinum 27, and this information was used to construct the curve in Figure 2-6. The thermoelectric power is about 9 μV/°C at 1300°C.

Walker, Ewing, and Miller (1962) experimented with this couple and found that the iridium, or some impurity in the iridium, strongly contaminated the Pt-10Rh alloy. Air exposure for 120 hours at 1380°C resulted in gross contamination and large changes in calibration.

(25-90)Pt-(0-70)Pd-(2-15)Rh vs (50-70)Pd-(25-40)Au-(0-15)Pt.   Thermocouples of alloys within these composition ranges were included in a patent by Schneider

(1967). Good oxidation resistance and stable operation to temperatures of about 1425°C were claimed. Base metal extension wires were also described which could be used at temperatures up to 1000°C and which matched the noble metal couples satisfactorily in the 500-1000°C range.

(25-90)Pt-(0-70)Pd-(2-15)Rh vs (52-68)Pd-(26-39)Au-(3-12)Pt.   Schneider (1967) stated that these thermocouples are especially suited for applications where stress and vibration are present and gave the composition range shown here, which is a restricted composition version of the couple in the preceding paragraph. The platinum is included in the negative alloy specifically for improvement of strength and resistance to vibration.

(25-90)Pt-(0-70)Pd-(2-15)Rh vs (60-70)Pd-(30-40)Au. Schneider also specified this composition range for applications such as furnace temperature measurements where stress and vibration are not critical and the platinum can be omitted from the negative conductor.

Pd-12.5Pt vs Au-46Pd.   An application of this combination for the temperature measurement of furnace waste gases was described by Dalton (1969). This thermocouple is available under the trade name[10] of Pallador II and has an output similar to that of Chromel-Alumel type (Type K) couples. Dalton found that Pallador II thermocouples in aluminum oxide insulators had a greatly improved service life at 1100°C compared with the Type K couple. Technical data from the manufacturer indicate that this couple can be used to 1200°C in nonreducing atmospheres, and the calibration curve is within ±1 per cent of the calibration for Chromel-Alumel over the 500 to 1200°C interval. An accuracy of ±2°C can be obtained using the Pallador II calibration curve supplied with each shipment.

Pd-14Pt-3Au vs Au-35Pd.   Thermocouples with this composition are available commercially as Platinel I,[11] and were first described in the literature during the early 1960's. Accinno and Schneider (1960, 1962) indicated that this couple was developed to improve upon the useful life of thermocouples used for temperature control and indication in turbojet engines. Chromel-Alumel thermocouples were in general use for this application, so Platinel I was tailored for a voltage-temperature curve which very closely matched that of Chromel-Alumel. A curve is given in Figure 2-7 from Zysk (1963) and Dalle Donne (1964). The thermoelectric power is near 44 μV/°C at 400°C, and 32 μV/°C at 1200°C. Zysk (1964) reported that extended use at 1200°C in air is possible, and at 1300°C for short periods of time. Accinno and Schneider (1962) reported the melting temperature of Pd-14Pt-3Au to be 1568°C; that of Au-35Pd was 1426°C.

Accinno and Schneider also conducted long period exposure tests and determined thermocouple drift. A couple was exposed to air at 1300°C for 1008 hours. A negative drift occurred at all calibration temperatures, ranging from -2°C or -0.5 per cent at 400°C to -1°C or about -0.08 per cent at 1300°C.

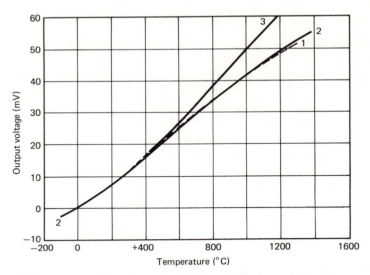

FIGURE 2-7   Multicomponent thermocouples containing platinum,
II.   1. Pd-14Pt-3Au vs Au-35Pd, Zysk (1963), Dalle Donne (1964).
2. Pd-31Pt-14Au vs Au-35Pd, Olsen and Freeze (1964).   3. Pt-5Rh
vs Au-46Pd-2Pt, Schulze (1939).

A table of drift data was also shown for a group of six thermocouples,
each of which was held for 2500 hours in air at a given exposure temperature.
These temperatures ranged from 400 to 1300°C. Calibrations were made and the
drift was shown for each couple at seven different calibration temperatures.
The greatest percentage drift was about –0.7 per cent, and in many instances
the change was considerably less.

Another type of test was also described in which a couple was subjected
to 20,500 temperature cycles between 100 and 1250°C, over a three month
period. The greatest percentage change was +1.2 per cent at 600°C. Changes of
–0.3 and –0.8 per cent were observed at 1000 and 1200°C, respectively.

Performance in an oxidizing atmosphere consisting of wet steam and carbon
dioxide, between 1000 and 1200°C, was described by Accinno and Schneider
(1960). After 1658 hours the maximum drift noted during recalibration was
–0.31 per cent. After an additional 1000 hours the drift was –0.44 per cent.
These values occurred at a calibration temperature of 400°C. Drift was 0.2 per
cent or less in magnitude in the 1000-1200°C region.

Dalle Donne (1964) discussed some differences observed between various
sets of calibration data and suggested that such differences may have been caused

by the annealing procedures which were used. The procedure used by Dalle Donne was to heat the thermocouple for 15 minutes in air at 1250°C. Caldwell (1962) described a treatment in which the positive wire was held at 1204°C for 15 minutes; the negative wire was held at 1093°C for 15 minutes.

Dalle Donne found that both thermocouple wires became embrittled after long exposure periods at 1100°C and above. The embrittlement was reported to be caused by grain growth. Crystalline growth was apparent in 0.51 mm wires used for drift tests. Zysk (1963) had described earlier thermocouple failures caused by fracture of the positive wire, apparently due to fatigue. The Pd-14Pt-3Au alloy is not as strong as Au-35Pd, and for applications where strength is a consideration, an alloy of somewhat different composition is used. The resulting thermocouple, Platinel II, is discussed under Pd-31Pt-14Au vs Au-35Pd.

Cormier and Claisse (1967) used Platinel, presumably Platinel I, for temperature measurements of titanium during quenching experiments. Couples were used for a few minutes at temperatures as high as 1100°C, after welding to titanium, with no significant diffusion at the weld. A 15-minute exposure at 1075°C gave no indications of alloying, although diffusion was observed to begin at approximately 1150°C.

Pd-31Pt-14Au vs Au-35Pd.   According to Accinno and Schneider (1962), this thermocouple was developed shortly after the development of Platinel I, and is called Platinel II.[11] The composition of the positive conductor was chosen because of its greater strength compared with the alloy used in Platinel I. As a result, Platinel II is superior for use where mechanical strength is needed to withstand loading, vibration, and fatigue. The voltage-temperature curve retains the similarity to Chromel-Alumel that Platinel I demonstrates, although there are some minor differences.

A calibration table for Platinel II was given by Zysk (1963). Shortly thereafter, Olsen and Freeze (1964) at the National Bureau of Standards developed a set of reference tables to meet growing interest in the thermocouple. The tables were based on work with 27 thermocouples, of wires drawn from three separate melts for each of the two alloys required.

All couples were annealed for 90 minutes in still air at 1316°C prior to calibration. A lengthy annealing period was considered necessary to eliminate large voltage changes that occur during the first hour of heating from the as-drawn condition. It had been found that during the first 0.5 hour of heating there was a large decrease in output, followed by partial recovery during the second 0.5 hour period. After prolonged heating the output became greater than the initial value.

Calibrations were made from which reference tables were constructed for the –100 to +1371°C interval. Some of this information is plotted in Figure 2-7. From these same reference tables, the thermoelectric power is 22 μV/°C at –100°C, 36 μV/°C at +100°C, and 45, 39, and 33 μV/°C at 500, 1000, and 1300°C, respectively.

It was concluded that the temperature of a Platinel II thermocouple, as determined directly from the tables, is within 0.5 per cent of the true temperature; this makes it unnecessary, for most work, to calibrate such couples and construct difference curves. It was pointed out that small changes in thermocouple outputs occurred during calibration, and these were of the same order of magnitude as calibration curve deviations from the tables, particularly in the 260-540°C range. This makes it difficult to apply difference curve corrections, although such procedure can still be used. The annealing procedure previously described was an effort to minimize these small changes as much as possible.

Direct use of the tables with an uncalibrated couple will provide for temperature measurements which are not in error by more than 1.7, 2.8, and 5.6°C at 260, 540, and 1370°C, respectively. By using calibrations at such temperatures as 260, 540, 820, and 1370°C, and constructing difference curves, the maximum error can be reduced from 5.6 to about 3.3°C.

The near equivalence of the voltage-temperature characteristic with that of Chromel-Alumel was also demonstrated. Between 540 and 1370°C the maximum output difference between Platinel II and Chromel-Alumel is the equivalent of only 10°C. The 1370°C maximum temperature is somewhat below the melting point of the negative alloy, which was reported by Accinno and Schneider (1962) to be 1426°C. The corresponding temperature for the positive alloy is 1570°C.

Freeze and Olsen (1962) conducted long period drift tests with five bare wire couples, in air, at various temperatures in the 871-1260°C interval. In general, drift was within 0.75 per cent for periods up to 1500 hours. At about 1200°C one couple was within that limit at 1000 hours, and only slightly exceeded 0.75 per cent after an additional 500 hours. At 1260°C, however, another thermocouple exceeded 0.75 per cent drift in about 100 hours.

Zysk (1963) described tests in air in which the couples were protected by aluminum oxide insulators between the cool region and a point slightly more than 1 cm distant from the hot junction. With this partial protection no objectionable drift occurred during 1000 hours at 1300°C. Tabulated data were shown for couples exposed for 1100 hours in still air at temperatures ranging from 600 through 1300°C. Each couple was calibrated at various times during the test and the extreme positive and negative microvolt deviations observed for each couple at each calibration temperature were shown. By using the reference tables of Olsen and Freeze (1964) and converting the microvolt deviations to percentages of the initial calibration temperature, it can be shown that drift was of the order of 1 per cent or less in magnitude in all cases.

Thermal cycling tests by Accinno and Schneider (1962) produced little change in calibration after 20,500 cycles between 100 and 1250°C in air. The maximum change of 0.33 per cent occurred at the calibration temperature of 600°C. Changes of –0.30 and –0.25 per cent were observed at 1000 and 1200°C, respectively. Freeze and Olsen (1962) subjected couples to thermal shock cycling, and found only slight changes after 600 such cycles.

According to Zysk (1963), simulated and actual flight tests, using Platinel II for temperature measurements in turboprop engines, resulted in long term performance at temperatures up to 1038°C. Probes were used for 2000 hours without observable mechanical or thermoelectric deterioration. Probes using Chromel-Alumel type couples had a 1000-hour life under similar conditions. Zysk recommended Platinel II for the 800-1300°C range where extended life in oxidizing atmospheres is required. More recent information by Baas and Mai (1971) on gas turbine temperature measurement applications indicated that Platinel type probes with bare wire junctions of 1.3 mm diameter wires have a useful life of 3000 hours for the 925 to 1050°C range and 1000 hours for the 1050 to 1200°C range.

Work with this thermocouple has also been done in a hydrogen atmosphere. Greenberg and Zysk (1963) reported that drift did not exceed ±0.75 per cent during 1000-hour exposures of couples at temperatures up to 1000°C. Results of Accinno reported by Zysk (1963) indicated that in an atmosphere of free flowing hydrogen at 1300°C, a lifetime of 200 to 400 hours can be expected. Incipient failure at 1300°C is caused by progressive, uniform, volatilization from the Au-35Pd wire. Zysk (1963) recommended that the precautions associated with use of Pt-10Rh vs Pt be followed when using Platinel II. In particular, for reducing atmospheres high purity aluminum oxide was suggested for insulators and protection tubes when temperatures above 1000°C are expected.

Olsen (1962) found that Platinel couples were among those capable of catalytic effects when exposed in mixtures of hydrogen, carbon monoxide, methane, and propane. Zysk (1963) indicated that no catalytic effects had been observed during simulated in-flight engine tests with Platinel probes, apparently because of a different gas mixture which was not subject to catalytic action by the presence of the thermocouple.

Pt-35Pd-13Au vs Au-35Pd.   This combination was among those described by Hill (1963) in a patent covering a number of different thin film couples. The films are prepared by firing, after painting the substrate with liquid suspensions of the metals in an organic liquid to form each thermocouple conductor.

Pt-5Rh vs Au-46Pd-2Pt, Pt-10Rh vs Au-30Pd-10Pt.   According to Zysk and Robertson (1971) the thermocouple with 5 per cent rhodium is the combination often referred to in Europe as the Alloy 40 vs Alloy 32 pair or as the Pallaplat[12] thermocouple. This is an older thermocouple that was reviewed by Schulze (1939) and from which the curve shown in Figure 2-7 was taken.[13] The thermoelectric power is 51.5 μV/°C in the 400-500°C region, rises to a maximum of 56 μV/°C near 800°C, and is 52.0 μV/°C in the 1100-1200°C range. The highest temperature for application work is about 1000 to 1200°C. The couple is highly resistant to many corrosive media, but must be protected from attack by carbon-bearing gases and from silicon. This type of couple, using an alloy of gold, palladium, and platinum, was originally described by Feussner (1927).

Rudnitskii (1959) discussed the very similar combination of Pt-10Rh vs Au-30Pd-10Pt and indicated that it is useful in fused salts and organic substances without additional protection. It may be used to about 1200°C, but is subject to drift at higher temperatures because of volatilization of the gold. An accuracy of ±1 to ±2°C was reported.

Pt-20Rh-10Ir vs Pt-20Rh.   Rudnitskii and Tyurin (1960) investigated a number of alloy pairs for possible high temperature application and included this combination. The results of melting point calibrations were shown, and a calibration table was also given, with voltages for temperatures between 0 and 1800°C in 100°C increments. The voltage-temperature curve is included in Figure 2-8. The thermoelectric power is 4.8 $\mu V/°C$ at 1000°C and 5.3 $\mu V/°C$ at 1600°C.

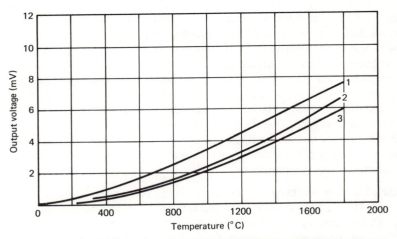

FIGURE 2-8   Multicomponent thermocouples containing platinum, III.   1. Pt-20Rh-10Ir vs Pt-20Rh;   2. Pt-20Rh-5Ru vs Pt-20Rh; 3. Pt-30Rh-10Pd vs Pt-20Rh: Rudnitskii and Tyurin (1960).

Drift tests were conducted in air after a two-hour preliminary anneal at 1200°C or at 1350°C. Regardless of the annealing temperature, there was first a positive drift for about 25 hours, followed by a negative drift. However, the positive drift was less in the case of the couple with the higher annealing temperature. The maximum drift magnitudes during a 100-hour test at 1800°C, after the 1350°C anneal, were +1.54 per cent at 25 hours and −1.07 per cent at

100 hours. The investigators ascribed the initial positive drift to recrystallization, and the subsequent negative drift to evaporation of the iridium and its diffusion into the negative conductor. It was remarked that the use of the ternary alloy for the positive conductor, rather than pure rhodium, had produced a less stable thermocouple.

Pt-20Rh-5Ru vs Pt-20Rh. This pair was also investigated by Rudnitskii and Tyurin (1960) and found to be less stable than Rh vs Pt-20Rh. A calibration was made at six temperatures by the melting point method, and a curve was also shown, which appears here in Figure 2-8. From this curve, the thermoelectric power was found to be 5 and 6 μV/°C at 1200 and 1600°C, respectively.

A drift test was conducted in air at 1800°C after a two-hour preliminary anneal at 1200°C. The drift was consistently negative, and amounted to -0.60, -5.24, and -10.80 per cent after 25, 50, and 100 hours, respectively. The relatively rapid drift was believed to be caused by evaporation of ruthenium and its diffusion into the negative conductor. From the later work by Israël (1966) with platinum-ruthenium alloys, it is likely that ruthenium oxidation was a significant factor here.

Pt-30Rh-10Pd vs Pt-20Rh. This thermocouple was evaluated by Rudnitskii and Tyurin in the same manner as was Pt-20Rh-5Ru vs Pt-20Rh, and was also found to be less stable than Rh vs Pt-20Rh. The calibration curve is shown in Figure 2-8, from which the thermoelectric power is approximately 4.5 and 5 μV/°C at 1200 and 1600°C, respectively. Drift at 25, 50, and 100 hours was -0.52, -0.71, and -2.76 per cent.

ELEMENTAL AND ALLOY SYSTEMS EXCLUDING PLATINUM

General Information

A few thermocouples utilizing palladium and other noble metals have been investigated, but the elements of most importance in this section are iridium and rhodium. Iridium and iridium-rhodium alloys have been utilized in several combinations, the best known being the Rh-40Ir vs Ir couple, which is also called the Feussner couple. Feussner (1933) studied the iridium-rhodium alloy system and recommended Rh-40Ir vs Ir for use at temperatures above the range of Pt-10Rh vs Pt. Since that time, the Ir-40Rh vs Ir and Ir-50Rh vs Ir couples have also been studied, and all three types are now used.

Iridium-rhodium versus iridium thermocouples can be used for extended periods in inert or mildly oxidizing atmospheres. Zysk and Robertson (1971) pointed out that thermocouples of these elements are the only known combinations that can be used without protection in air up to 2000°C. Even for these exceptional metals only 16 to 24 hours of operation is permissible at such temperatures. Vacuum applications have also been reported.

These metals are relatively ductile as received from the manufacturer, but become quite brittle because of grain growth during heating. Bending that occurs during normal handling is sometimes sufficient to cold work the wires to a brittle state. Another factor to consider is catalyticity. The catalytic effect of iridium-rhodium alloy couples has been described for several high temperature atmosphere conditions. The effect is not generally as pronounced as with platinum-rhodium alloys, but it is a very definite condition to contend with.

The properties of iridium coupled with alloys of 40 to 60 per cent rhodium have been discussed by Carter (1950), Steven (1956), Caldwell (1962), Zysk (1964), and Zysk and Robertson (1971). Because of the similarities, and the relatively minor advantages and disadvantages of these thermocouples, it is difficult to make a general judgment concerning over-all superiority of any one system. The Rh-40Ir vs Ir pair may be slightly less suitable than its competitors, although there is more application information available for it than for the newer alternate choices.

Zysk remarked that the iridium member is the weak link in these systems. The possibility of adding rhodium to the pure element to improve the oxidation resistance was suggested, although the magnitude of the thermoelectric power relative to the iridium-rhodium positive alloy would be decreased. Such a couple had been mentioned by Ihnat and Hagel (1960) and consisted of an all-alloy pair analogous with the all-alloy platinum-rhodium thermocouples.

Heat treatment has little effect on stability and drift in iridium. No significant changes were noted by Walker, et al. (1962) between as-drawn wires and wires heated at 1900°C for one minute. Exposure tests of 24 hours at 1400°C in argon revealed no excessive changes of output. Blackburn and Caldwell (1962) reported that iridium and iridium alloy wires are somewhat discolored after high temperature heating in air, as well as being embrittled from recrystallization. Nevertheless, there is little change in the thermoelectric properties.

Thermocouples

Ag vs Pd.   Thermocouples using silver and palladium were chosen by Barber and Pemberton (1955) for use as secondary standards at the National Physical Laboratory in England. It was believed that the use of pure elements rather than alloys would result in improved homogeneity and lack of drift if protection from contamination was provided. The possibilities of diffusion at the hot junction and oxidation of the metals were pointed out as disadvantages. Oxidation was not found to be a problem in the range of application.

Wires of 0.5 mm diameter were prepared by annealing the silver at about 700°C and the palladium at about 1400°C. Thermocouple wires were enclosed in twin bore silica tubes and hot junctions were prepared by fusing the wires. The thermocouples were calibrated between 200 and 600°C, from which a table was prepared of voltage versus temperature in 50°C intervals. A curve from these data is plotted in Figure 2-9. The thermoelectric power was said to be 18 $\mu V/°C$ at 200°C and 34 $\mu V/°C$ at 600°C.

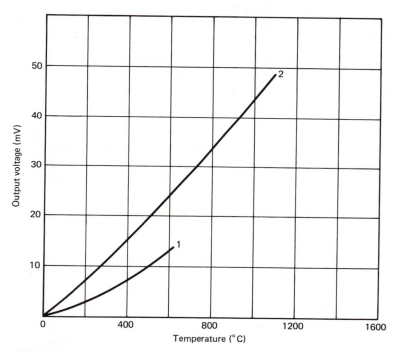

FIGURE 2-9   Thermocouples with palladium.   1. Ag vs Pd, Barber and Pemberton (1955).   2. Rh vs Pd, Ihnat and Hagel (1960).

The investigators stated that a calibration accuracy of ±0.1°C could be established and maintained for a considerable period. However, there were wire differences from batch to batch, and individual calibration of each couple was required. The units could be used up to 600°C in oxidizing atmospheres and were employed as secondary standards for base metal thermocouple calibrations. Couples were used for 30 to 60 tests per year for three to five years in the laboratory, but Ag vs Pd was not recommended as a couple for other applications. The wires were not mechanically strong and it was advised that bending be kept to a minimum.

As a result of heating in air a bluish-black discoloration of the palladium wire was observed. Experiments after 70 hours at 600°C detected no changes produced by this oxidation.

Bianchi and coworkers (1966) used a number of thermocouples calibrated by comparison with an Ag vs Pd secondary standard. An accuracy of ±0.3°C was reported for the resulting calibrations using this method.

Pd vs Pd-(10-60)Au. Thermocouples in this range of compositions were described in a patent by Hill (1963) for thin film alloy couples. The films are prepared by firing, after painting the substrate with a suspension of the metals in an organic liquid to form each conductor. The preparation of a Pd vs Au-40Pd couple was described.

Rh vs Pd. This was discussed in three reports: Ihnat (1959), Clark and Hagel (1960), and Ihnat and Hagel (1960). A voltage-temperature curve from Ihnat and Hagel is given in Figure 2-9. The thermoelectric power is about 51 μV/°C in the 800-1100°C region. Rh vs Pd and a number of other combinations were evaluated for use in a jet engine temperature measurement application.

A test was conducted in a chamber where the combustion atmosphere in a jet engine was simulated. Fuel oil containing 0.4 per cent sulphur was burned and temperatures ranged from 1100 to 1370°C. Drift from the initial calibration was determined by comparison with a Pt-10Rh vs Pt thermocouple at intervals of 25 or 50 hours. A somewhat irregular drift was observed when checked at 1093°C, which remained within approximately ±0.1 per cent during 450 hours of exposure at temperatures not exceeding 1316°C. Drift was within the limits of ±0.5 per cent after 600 hours.

The drift performance of this couple compared favorably with that of Pt-15Ir vs Pd, which drifted about 0.35 per cent from its initial output during 400 hours of exposure. Rh vs Pd and Pt-13Rh vs Pt were considered to be the least prone to drift of all the thermocouples tested.

Ir-10Rh vs Ir. This thermocouple was first discussed by Hoffmann (1909). According to Rohn (1927), the output was about 10 mV at the maximum temperature of 2300°C. Stauss (1941) stated that it was used for temperatures above the limit of Pt-10Rh vs Pt, but that it had the serious disadvantages of insensitivity, drift, and brittleness. A voltage-temperature curve is plotted in Figure 2-10 from more recent tabulated data by Blackburn and Caldwell (1964). The thermoelectric power is low in the entire 0 to 2200°C temperature range; the value varies from less than 2 to about 3 μV/°C. The output at 2204°C is only 5.45 mV, a considerably lower figure than that given in the early work of Hoffmann.

Ir-25Rh vs Ir. Thermocouples of this composition are commercially available in Germany, according to Wisely (1955). A curve from Blackburn and Caldwell (1964) is given in Figure 2-10; the thermoelectric power S is greatest at 540°C, where it is approximately 5.3 μV/°C. At 2200°C, S is 4.1 μV/°C, and is very low in the near room temperature region. At 38°C the output is 105 μV.

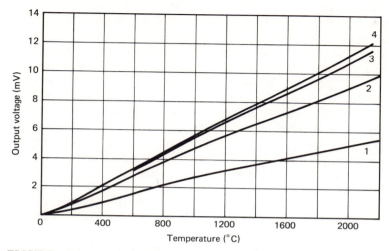

FIGURE 2-10   Iridium-rhodium alloy thermocouples, I.   1. Ir-
10Rh vs Ir;   2. Ir-25Rh vs Ir;   3. Ir-40Rh vs Ir;   4. Ir-50Rh vs
Ir: Blackburn and Caldwell (1964).

Ir-40Rh vs Ir.   Reference tables for this couple were prepared by Blackburn and
Caldwell (1964) and a voltage-temperature curve from this work is shown in
Figure 2-10. The thermoelectric power is 5.5 μV/°C at 1000°C and 5.8 μV/°C at
2150°C.

The wires were heat treated before calibration by electrically heating each
metal for one minute in air at about 200°C below the melting point. Four
thermocouples were prepared from wires supplied by two different manufacturers, and were calibrated to a temperature of 2150°C. Reference tables were then
prepared from the averaged data to cover the 0 to 2150°C range.

The investigators concluded from their study that an accuracy of ±22°C
can be attained if the temperature corresponding to a thermocouple voltage is
determined directly from the tables. It was suggested that accuracy can be increased by means of thermocouple calibration plus construction of a difference
curve with respect to the tables. This type of thermocouple is commercially
available, and extension wires of copper and Type 304 stainless steel can also
be purchased.

The melting point of iridium is 2447°C. Caldwell (1962) gave a calculated
temperature of 2250°C for the solidus point of Ir-40Rh. It was recommended
that the maximum application temperature be kept 60°C or more below the
alloy solidus point. This gives a maximum operating temperature of about

2190°C, and is the highest operating temperature of the three iridium-rhodium couples using iridium with 40, 50, and 60 per cent rhodium alloys.

Carter (1950) preferred this couple because the voltage drift that occurs is small when iridium volatilizes preferentially from the alloy. Steven (1956) discussed this condition, which is equivalent to adding rhodium to the alloy. It is known that the curve of voltage versus rhodium content at constant temperature shows a maximum at 50 per cent rhodium. Therefore, as iridium volatilizes from the 40 per cent rhodium alloy, the voltage first rises slightly and then declines to its original value as the alloy composition progressively increases in rhodium content from 40 per cent through about 60 per cent. This behavior was cited as an advantage of the Ir-40Rh vs Ir thermocouple since the net voltage change is small and therefore the drift is small over a 20 per cent change in composition from 40 to 60 per cent rhodium, an unusual situation. With higher rhodium content the drift becomes more pronounced.

Moeller, et al. (1968) described applications where Ir-40Rh vs Ir thermocouples were used, and recommended that for field work the wires should be handled in the as-drawn rather than the heat treated state because of the brittleness problem. Calibration was said to be little influenced by being in the as-drawn state. Successful use for gas temperature measurements was reported at the Langley Research Center; bare wire couples were exposed to air and to combustion products of propane fuel, at 1650°C, 200 atmospheres, and about 10 meters/sec flow velocity, for brief periods up to 20 minutes. Recalibrations after such exposures were within the ±1 per cent accuracy of the instrumentation system.

Gaylord and Compton (1968) exposed couples to slowly moving air for 50 to 100 hours at 1316°C. All specimens eventually failed due to oxidation of the iridium conductor. The alloyed conductor showed little iridium loss. Times to failure varied from 24 to 104 hours. An alternating positive and negative drift was experienced prior to failure with extremes of up to +18.3 and –57.2°C. Similar tests were conducted in argon at 1371°C for 74 hours with little visible thermocouple deterioration. Some drift was shown, mostly positive; the maximum apparent temperature change was about 16°C.

Ir-50Rh vs Ir.   Blackburn and Caldwell (1964) calibrated three thermocouples from two different lots of wire, and prepared reference tables, using the method described under Ir-40Rh vs Ir. The tables cover the 0-2150°C range in 10°C increments. The voltage-temperature curve appears in Figure 2-10, and the thermoelectric power is 5.7 and 6.2 $\mu V/°C$ at 1000 and 2150°C, respectively. The calculated solidus temperature, from Caldwell (1962), is 2202°C. Using the rule that the maximum use temperature be at least 60°C lower, a maximum operating temperature of about 2140°C results. This is somewhat less than the 2190°C maximum temperature for the Ir-40Rh vs Ir couple. The output, on the other hand, is slightly higher for Ir-50Rh vs Ir as shown in Figure 2-10. The 50 per cent rhodium alloy has the highest thermoelectric power of all the Ir-Rh alloys. Also, output is not critically dependent on the alloy composition. A positive or negative deviation from the nominal 50 per cent rhodium content results in only a small change of output at any given temperature.

Blackburn and Caldwell concluded that an accuracy of ±22°C can be attained if the temperature corresponding to a thermocouple voltage is determined directly from the reference tables. Accuracy can be increased by thermocouple calibration plus construction of a difference curve. Wires of these compositions are commercially available, and extension wires of copper and Type 304 stainless steel can be used.

Walker, Ewing, and Miller (1962; 1965, p. 601) included this couple in their studies of thermocouple drift and made some interesting comparisons with platinum-rhodium alloy thermocouples. Experiments in environments of air, argon, and vacuum were made using couples enclosed in low iron content aluminum oxide insulators, and with a given furnace firing temperature maintained for 120 hours. Drift was determined in terms of the apparent change in thermocouple temperature observed at an 860°C calibration temperature. One sample from the extensive results is given in Table 2-7. The experiments were conducted with 0.51 mm diameter wires. The insulators were from two manufacturers and are shown here as Types 1 and 2. These and other results led Walker and coworkers to the conclusion that the Ir-50Rh vs Ir couple appeared more reliable than any of the platinum-rhodium alloy thermocouples, in either argon or vacuum at high temperatures, and where there was either minimum or gross iron contamination. This was also true in air, but only where gross iron contamination existed. At lower temperatures, or in air, when minimum iron contamination from the insulators occurred, performance was not as satisfactory for Ir-50Rh vs Ir as for platinum-rhodium alloy couples. This condition was attributed to small internal changes in the thermocouple conductors which decrease the reliability of iridium-rhodium thermocouples.

The drift experiments described above were conducted with Ir-50Rh vs Ir. Although it would seem that similar results should occur with Ir-40Rh vs Ir and Rh-40Ir vs Ir, Walker, et al., found evidence that these other alloys were more prone to drift, during tests in argon at 1600°C. Because of this condition, the remaining tests were made with Ir-50Rh alloy wires, and with thermocouples using that alloy.

Stanforth (1962) used the Ir-50Rh vs Ir thermocouple to measure combustion gas temperatures in the 927 to 1593°C temperature range, where catalytic effects were observed. A flame-sprayed aluminum oxide coating on the wires appeared to prevent such an effect. Catalysis in the presence of iridium and iridium-rhodium alloys has also been reported by Olsen (1962) and Alvermann and Stottmann (1964).

TABLE 2-7

Drift of Ir-50Rh vs Ir Thermocouples during Exposure for 120 Hours in Al_2O_3 Insulators

Firing Temperature (°C)	Drift in Celsius Degrees[a] in the Environment Listed with Insulator Type 1 or 2[b]					
	Air		Argon		Vacuum	
	1	2	1	2	1	2
1200	+2	+2	+2	+2	+3	+3
1380	+2	+2	0	-1	+3	+2
1600	+4	- -	-2	- -	+4	-9

[a] Measured at the calibration temperature of 860°C.
[b] The two types of aluminum oxide insulators were supplied by different manufacturers.

Rh-40Ir vs Ir. This was the combination chosen by Feussner (1933) after studying the iridium-rhodium alloy system. According to Schulze (1939), the fact that alloys of more than 10 per cent rhodium could be worked was unexpected, and was taken advantage of in selecting the Rh-40Ir alloy for thermocouple use. This couple is often referred to as the Feussner thermocouple.

From the late 1940's into the 1950 decade, work was done by Droms and Dahl (1955) on improving the quality of the wires and developing practical probes for gas temperature measurement applications where platinum-rhodium alloy couples could not be used. Feussner, Droms and Dahl, and also Rudnitskii and Tyurin (1960), gave calibration data; Blackburn and Caldwell (1962) prepared the reference tables that are now used with the couple. The latter work was done by the procedure described under Ir-40Rh vs Ir. Eight thermocouples made from three different lots of wire were calibrated.

The tables cover the 0 to 2100°C temperature range in 10°C increments. From information by Caldwell (1962), the computed solidus temperature of the alloy is 2153°C and the maximum use temperature is about 2090°C. A curve from the work of Blackburn and Caldwell is shown in Figure 2-11. The thermoelectric power S is nearly constant from about 300°C up to 1760°C; S is 5.4 $\mu V/°C$ at 300°C and 5.9 $\mu V/°C$ at 1760°C. The value of S increases somewhat at higher temperatures, reaching 6.6 $\mu V/°C$ at 2100°C. Blackburn and Caldwell (1964) later stated that an accuracy of ±22°C can be reached by direct determination of temperature from the tables, using the thermocouple voltage. Accuracy can be improved by making a thermocouple calibration and then constructing a difference curve with respect to the tables.

Troy and Steven (1950) observed good stability characteristics for this couple during an investigation of noble metal combinations. Droms and Dahl (1955) applied Rh-40Ir vs Ir to measurements in a combustion tunnel and of afterburner discharges, following their development and improvement program. Kostkowski and Burns (1963) made estimates of drift in inert or mildly oxidizing atmospheres, based on data in the literature and experience at the National Bureau of Standards. A change of 10°C in calibration was estimated for ten hours at 2000°C. Kuether and Lachman (1960) cycled a couple between room temperature and 1427°C ten times or more in vacuum and accumulated a total exposure time of over 20 hours. The wires were presumably unprotected. A negative drift of −11.7°C was observed, measured at 1093°C.

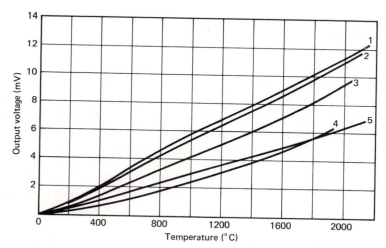

FIGURE 2-11   Iridium-rhodium alloy thermocouples, II.   1. Ir-50Rh vs Ir, Blackburn and Caldwell (1964).   2. Rh-40Ir vs Ir, Blackburn and Caldwell (1962).   3. Rh-25Ir vs Ir;   4. Rh-10Ir vs Ir;   5. Ir-50Rh vs Ir-10Rh: Blackburn and Caldwell (1964).

Rudnitskii and Tyurin (1960) tested several types of thermocouples and found that this combination gave good results. Tests were made using aluminum oxide insulators containing 1 per cent titanium dioxide, and also with pure beryllium oxide. Drift of 0.6 per cent occurred during the first 25 hours of exposure at 1800°C, but this was attributed to recrystallization in the wires. Subsequent drift during an additional 75 hours did not exceed ±0.3 per cent.

Bennett (1961, p. 132) found that aluminum oxide insulators containing 0.2 per cent silicon dioxide contaminated a couple exposed to a reducing atmosphere at 1400°C. The silicon dioxide was reduced to silicon which then combined with the rhodium in the grain boundaries. Aleksakhin and co-authors (1967) have made a detailed review of work on this and other thermocouples.

Rh-30Ir vs Ir.   This thermocouple was shown with others in a set of voltage-temperature curves given by Stadnyk and Samsonov (1964) in a review article. The output was quite nonlinear and exceeded that of Rh-40Ir vs Ir at temperatures above about 1400°C. Such a characteristic is not typical of the results of Blackburn and Caldwell (1962, 1964); Rh-40Ir and Rh-25Ir show smooth, near linear curves relative to iridium. One would expect Rh-30Ir vs Ir to have an intermediate output between the curves for these two couples and to be without significant nonlinearities.

Rh-25Ir vs Ir, Rh-10Ir vs Ir.   Data from Blackburn and Caldwell (1964) are included in Figure 2-11. It was suggested that these couples may find use in special applications. Their outputs are low relative to the better known combinations of iridium-rhodium versus iridium. At 1600°C the thermoelectric power S is 5.4 and 4.5 $\mu V/°C$, for Rh-25Ir vs Ir and Rh-10Ir vs Ir, respectively. S is lower below this temperature in both cases.

Ir-50Rh vs Ir-10Rh.   The voltage-temperature curve in Figure 2-11 is constructed from data on Ir-50Rh vs Ir and Ir-10Rh vs Ir given by Blackburn and Caldwell (1964). Ihnat and Hagel (1960) evaluated such a couple with a number of other types. A 600-hour test was conducted in a chamber where the combustion atmosphere in a jet engine was simulated. Fuel oil with 0.4 per cent sulphur content was burned; the maximum temperature was 1316°C. The thermocouple drift was less than ±0.5 per cent during the 600-hour period and performance was better than that of a Rh-40Ir vs Ir thermocouple. No other details were given.

Zysk (1964) commented on the possibilities of an all-alloy iridium-rhodium thermocouple. Although the output is lower than that of a similar couple with pure iridium as one conductor, the oxidation resistance of the alloys is significantly improved over that of iridium, leading to longer life.

Ir vs Ir-10Ru.   Schulze (1939) reviewed this combination, giving a calibration table from 1000 to 1800°C, from which the curve in Figure 2-12 was plotted. The thermoelectric power S is very low, being 1.5 $\mu V/°C$ in the 1700-1800°C range.[14] The pair was first described by Hoffmann (1909) who made a calibration by the melting point method using gold, palladium, and platinum. The wires were brittle, and at high temperatures in air the oxidation of the ruthenium was rapid, so that frequent recalibration was required.

Ir-10Rh vs Ir-10Ru, Rh-40Ir vs Ir-10Ru.   Both of these couples were also studied by Hoffmann. Although used up to about 2000°C in air, the loss of ruthenium by oxidation was too rapid for good stability. Müller (1930) utilized the Ir-10Rh vs Ir-10Ru couple to determine the melting point of chromium, after calibrating by comparison with a platinum-rhodium versus platinum thermocouple to 1600°C and then making a platinum point calibration. Müller's curve is given in Figure 2-12. The thermoelectric power is about 6.3 $\mu V/°C$ in the 600-1200°C region. The wires were found to be very brittle and broke easily. Very careful handling was required.

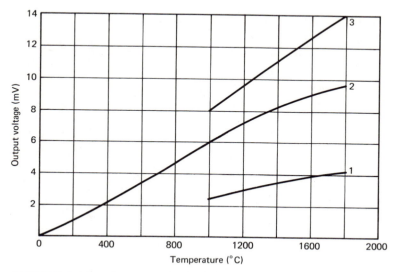

FIGURE 2-12   Thermocouples with iridium, rhodium, and ruthenium.   1. Ir vs Ir-10Ru, Schulze (1939).   2. Ir-10Rh vs Ir-10Ru, Müller (1930).   3. Rh-40Ir vs Ir-10Ru, Schulze (1939).

A brief discussion of Rh-40Ir vs Ir-10Ru was given by Aleksakin, et al. (1967). The voltage-temperature curve in Figure 2-12 was plotted using a table from Schulze (1939), who also gave a value of 7.6 $\mu V/°C$ for thermoelectric power in the 1200 to 1400°C region.

THERMOELECTRIC DATA ON MISCELLANEOUS NOBLE METALS

Some additional voltage-temperature curves for noble metals relative to platinum are given in Figures 2-13 through 2-16. These are included here because little or no thermocouple performance data were available for such combinations, and thus they did not appear in the preceding sections. Most of these materials are used as conductors in thermocouples with metals other than elemental platinum. Pt-10Rh vs Pt and Pt-13Rh vs Pt are included in Figure 2-16 for comparative purposes.

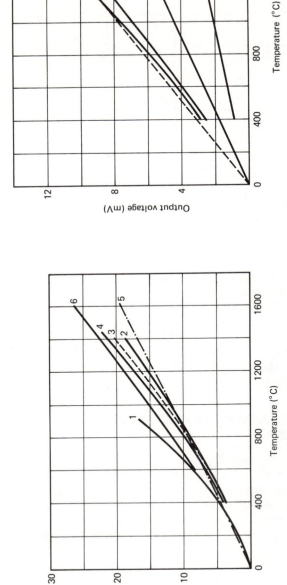

FIGURE 2-13   Additional thermoelectric voltage data on noble metals, I.   1. Ag vs Pt, Roeser and Wensel (1941, p. 1293). 2. Pt-20Rh vs Pt;   3. Pt-30Rh vs Pt;   4. Pt-40Rh vs Pt: Zysk   (1962).   5. Pt-2Ru vs Pt; 6. Pt-4Ru vs Pt: Israël (1966).

FIGURE 2-14   Additional thermoelectric voltage data on noble metals, II.   1. Pt-1Rh vs Pt;   2. Pt-5Rh vs Pt; 3. Pt-6Rh vs Pt: Zysk (1962). 4. Pt-0.5Ru vs Pt; 5. Pt-1Ru vs Pt: Israël (1966).

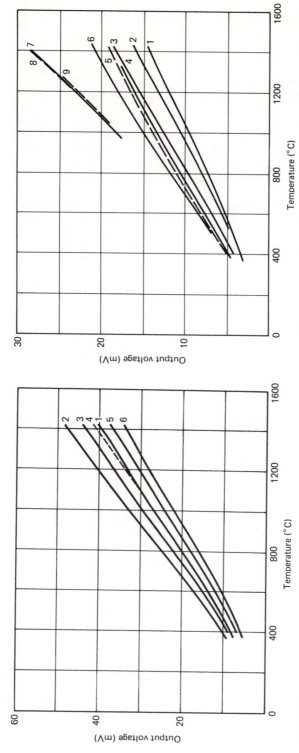

FIGURE 2-16 Additional thermoelectric voltage data on noble metals, IV. 1. Pt-10Rh vs Pt; 2. Pt-13Rh vs Pt; 3. Pt-38Pd-5Rh vs Pt; 4. Pd-38Pt-5Rh vs Pt; 5. Pd-33Pt-6Rh vs Pt; 6. Pd-32Pt-8Rh vs Pt: Schneider (1967). 7. Ir-40Rh vs Pt; 8. Ir-50Rh vs Pt; 9. Rh-40Ir vs Pt: Zysk (1962).

FIGURE 2-15 Additional thermoelectric voltage data on noble metals, III. 1. Pt vs Pd-30Au; 2. Pt vs Pd-40Au; 3. Pt vs Pd-39Au-2. 5Pt; 4. Pt vs Pd-38.2Au-4.5Pt; 5. Pt vs Pd-37Au-7.5Pt; 6. Pt vs Pd-36Au-10Pt: Schneider (1967).

FOOTNOTES

1. All references are located at the end of the book.
2. The composition was not stated but presumably was the Pt-10Rh vs Pt or Pt-13Rh vs Pt type of couple.
3. The platinum melting point of $1772°C$ was estimated by Sosman to be at $1755 ±5°C$. For work done in 1910 this was a considerable achievement. Lower temperatures on Sosman's scale agree remarkably well with the present temperature scale for the Pt-10Rh vs Pt thermocouple.
4. The black oxide film which appears on the alloy wire under certain conditions is discussed in a later paragraph.
5. A subcommittee of the British Iron and Steel Research Association.
6. Analysis showed the presence of rhodium in the platinum wire, but the text implies that there was no increase detected compared with unused wire.
7. Insulators from two sources were used; results for the best performing thermocouples are given here.
8. Manufactured by Johnson, Matthey and Co., Ltd.
9. At this temperature Mach 0.1 corresponds to an air stream velocity of 74.3 meters per second.
10. Manufactured by Johnson, Matthey and Co., Ltd.
11. Manufactured by Engelhard Minerals and Chemicals Corp.
12. Supplied by W.C. Heraeus GmbH, Hanau, West Germany.
13. Later data by Zysk and Robertson based on one test sample showed a somewhat lower output, suggesting minor differences in alloy composition. Schulze gave no compositions in his review.
14. Schulze also described an output of 10 mV at $2300°C$, however. Further, an accompanying voltage-temperature curve was not in full agreement with the tabulated information. These discrepancies may have been due to use of data from different sources.

3

Thermocouples of Mixed Noble and Base Metals

ELEMENTAL AND BINARY SYSTEMS

General Information

The large thermoelectric powers of some of the alloys of platinum with base metals encouraged early investigation of these metals for thermocouple use at high temperatures. It was found, however, that in many cases these alloys have a marked tendency to recrystallize and become embrittled during high temperature exposure. Further, elements such as iron, nickel, cobalt, or chromium tend to lower the melting point of the alloy, compared with pure platinum; they can also oxidize or volatilize from the alloy, depending on the type of atmosphere. These characteristics restrict the temperature range and other conditions for thermocouple use.

In a few instances, where a small amount of a high melting point element such as molybdenum, rhenium, or tungsten is added to the platinum, the melting point of the alloy is raised above that of pure platinum. In this case, temperatures above the pure platinum melting point may be reached before volatilization from the solid alloy becomes significant. At one time this property was important enough to encourage limited use of the resulting high-output high-temperature couple, even when oxidation of the base metal component occurred and embrittlement problems existed.

Schulze (1939)[1] described efforts to decrease embrittlement by adding small amounts of a third element, which inhibited recrystallization without seriously decreasing the thermoelectric power. Nishimura (1958), in his patent, mentioned the use of a third element to improve the workability of certain platinum base binary alloys. Examples of the latter are alloys of platinum with more than 7 or 8 per cent tungsten; Carter (1950) reported that such metals are quite difficult to prepare in wire form.

Improvements in the sensitivity of measurement instruments reduced the incentive to develop high output couples of noble-base alloys. There was renewed interest in certain alloys by 1950, however, because of the development of nuclear reactors. The interest in nuclear applications arises from the fact that some of the elements have small neutron collision cross sections. Thermocouples of such elements show a low rate of transmutation in a nuclear environment; this results in a slow rate of contamination of the thermocouple due to buildup of new and unwanted elements in the original alloy. The platinum-molybdenum alloys have been particularly interesting in this respect.

A number of applications have been found for non-platinum noble metal alloys with base metals. These include cryogenic temperature measurement in the 1 K region, and high temperature work at and above the melting point of platinum. For cryogenic use, a very small amount of a selected base element is added to the noble metal, usually gold, to produce a dilute alloy with a high thermoelectric power in the 10 to 20 K temperature region. High temperature applications have involved the use of alloys with large amounts of either the noble or the base metal. The solute element is usually the component with the lower melting point, and the alloying of the two elements improves the chemical and mechanical properties of the thermocouple at high temperatures.

Elemental and Binary Alloy Thermocouples Containing Platinum

Pt vs Cu-0.0016at.In.   Platinum versus copper-indium was selected by Pearson (1954) for a cryogenic application in the 4.2 to 20 K range, after investigating a number of dilute alloys of copper that had the same general voltage-temperature characteristics. The couple was used with the reference junction in a liquid helium bath; it was indicated, however, that the thermoelectric power dropped at higher temperatures and that the couple would be suitable for use with a higher reference junction temperature. The thermoelectric power was 5 μV/K and 10 μV/K at 5 and 10 K, respectively.

Pt vs Fe.   According to Zysk (1960), and Hunt (1964), a very early use of this thermocouple for measurement purposes was reported by C.S.M. Pouillet in 1836. His application consisted of measuring furnace temperatures after calibrating a platinum-iron couple by means of a gas thermometer. The couple consisted of a platinum wire in the barrel of a gun, which acted as a sheath. The wire was insulated with magnesium oxide or asbestos, and contacted the gun metal at the breech, forming the hot junction.

No present day references for Pt vs Fe have been found. Such a thermocouple would be a poor performer at high temperatures because of platinum contamination by the iron.

Pt vs Au-0.02at.Fe. Information on this pair was given by Sparks and Hall (1969). A voltage-temperature curve is shown in Figure 3-1 based on a 0 K reference junction temperature. The thermoelectric power reaches a maximum in the 10 to 15 K region, with a magnitude of a little over 15 μV/K. This couple has the disadvantage of high platinum thermal conductivity, which is usually undesirable for cryogenic work.

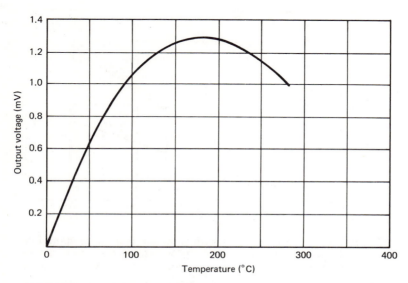

FIGURE 3-1 Voltage-temperature data for Pt vs Au-0.02at.Fe from Sparks and Hall (1969). The reference junction temperature is at 0 K.

Pt-1Mo vs Pt, Pt-3Mo vs Pt, Pt-3.5Mo vs Pt, Pt-4Mo vs Pt. Voltage-temperature curves are shown in Figure 3-2 from the patent by Nishimura (1958). These alloys were reported to be somewhat unsatisfactory in terms of tensile strength, electrical resistivity, and quality for wire drawing. The Pt-3Mo vs Pt couple was used continuously for three months at 1300°C with no noticeable output drift; the type of environment was not specified but was apparently an oxidizing atmosphere.[2]

Darling, et al. (1970, 1971) described rapid reaction of Pt-1Mo wire in contact with powdered aluminum oxide at 1600°C. An atmosphere of argon with about 50 ppm impurities was present. Reducing the oxygen impurity level by gettering the argon or operating in vacuum did not eliminate reactivity.

Pt-5Mo vs Pt. The remarks of Nishimura concerning the platinum alloys described above also apply here. Zysk (1964) presented some tabulated data for

Pt-5Mo versus Pt 27, the National Bureau of Standards' platinum reference standard at that time. A condensed version of this information is plotted in Figure 3-2. The thermoelectric power ranges from about 35 to 40 μV/°C in the 800-1600°C temperature range. Data by Nishimura (1958; 1959, p. 425) for Pt-5Mo vs Pt showed a lower output than that of Pt-4Mo vs Pt. This is unusual, since curves of output versus molybdenum content at a fixed temperature, such as the one shown by Schulze (1939), have no maximum in the 4 to 6 per cent molybdenum content range.

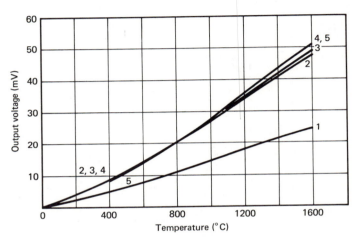

FIGURE 3-2   Thermocouples with platinum and platinum-molybdenum alloys.   1. Pt-1Mo vs Pt;   2. Pt-3Mo vs Pt;   3. Pt-3.5Mo vs Pt;   4. Pt-4Mo vs Pt: Nishimura (1958).   5. Pt-5Mo vs Pt, Zysk (1964).

Zysk recommended that this combination not be used in air above 600°C. Other information by Reichardt (1963) for couples of similar composition indicates that drift can occur if the slightest amount of oxygen is present in an otherwise inert atmosphere. Bennett, Hemphill, Rainey, and Keilholtz (1966) studied out-of-pile thermocouple behavior of these metals for a nuclear application. The wires were in contact with graphite and there were other contaminating conditions present.

Pt-5Mo vs Pt-0.1Mo.   A curve from work by Reichardt (1963) is shown in Figure 3-3; a bare wire couple of this type was tested for 1000 hours at 1300°C in commercially pure argon, and a 15 per cent drift in output was observed, probably caused by trace amounts of oxygen. Another 1000-hour test in argon, but with the couple inserted in a graphite block, produced no detectable drift. The extreme sensitivity to oxidation appeared to be inhibited by the presence of graphite. A test for 400 hours at 1500°C, in vacuum and with graphite present, also resulted in no detectable drift.

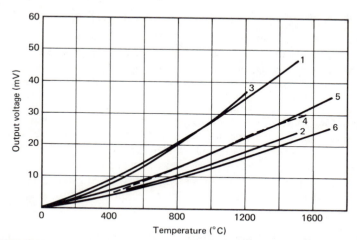

FIGURE 3-3   Thermocouples with molybdenum, platinum-molyb-denum, and other metals.   1. Pt-5Mo vs Pt-0.1Mo, Reichardt (1963).   2. Pt-5Mo vs Pt-1Mo, Reichardt (1963).   3. Mo vs Pt, Roeser and Wensel (1941, p. 1293).   4. Mo vs Pt-10Rh, Van Liempt (1929).   5. Pt-4Mo vs Pt-13Rh, Nishimura (1958).   6. Pt-5Mo vs Pt-13Rh, Nishimura (1958).

Couples of this type are of interest in nuclear applications because of the low neutron cross sections of platinum and molybdenum. Such use was described by J. Lagedrost and J. Stang in the report by Cunningham and Goldthwaite (1966). The couples were installed in test capsules for the 1540-1760°C range in a nuclear environment. In this case, however, substantial drift was observed. The investigators suggested that a contamination process was responsible.

Tseng, Robertson, and Zysk (1968) calibrated thermocouples from five different lots of wire and developed tables of output versus temperature in 1°C increments over the 0-1600°C range. The thermoelectric power is 34 and 36 μV/°C at 1000 and 1600°C, respectively. The tables indicate that the output is a few per cent lower than that shown by Reichardt. The annealing process to prepare the wires for calibration consisted of exposure to a temperature of 900°C for the Pt-0.1Mo and 1400°C for the Pt-5Mo; the exposure time was about seven minutes in both cases. An atmosphere of nitrogen with 7 per cent hydrogen was used.

Pt-5Mo vs Pt-1Mo.   Reichardt also suggested that a Pt-1Mo alloy may be preferable to the Pt-0.1Mo alloy used with Pt-5Mo. A voltage-temperature curve is included in Figure 3-3. Above 1000°C the curve appears to be nearly linear with a thermoelectric power of about 20 μV/°C.

Mo vs Pt.   A voltage-temperature curve is plotted in Figure 3-3 from a table by Roeser and Wensel (1941, p. 1293). Hawkings (1964) reviewed a number of reports on neutron flux monitors and thermocouples for in-core reactor measurements, and mentioned a molybdenum-platinum couple. The application was attributed to Robertson (1956).

Mo vs Pt-10Rh.   According to Troy and Steven (1950), this was one of the first thermocouples that combined noble and base metals for high temperature measurement, and was the result of work by Van Liempt (1929). Data on the voltage-temperature characteristic are plotted in Figure 3-3. The upper limit of usefulness was given as 1830°C. Van Liempt substituted molybdenum for platinum in the Pt-10Rh vs Pt couple in an effort to improve performance in reducing atmospheres.

Pt-4Mo vs Pt-13Rh, Pt-5Mo vs Pt-13Rh.   The output curves from the patent of Nishimura (1958) are included for both of these couples in Figure 3-3. The thermoelectric power of Pt-4Mo vs Pt-13Rh is about 25 μV/°C from the figure; the second combination is less sensitive, about 19 μV/°C. Both values apply to the 1200 to 1400°C region. It is evident that the thermoelectric power declines at lower temperatures. The voltage-temperature curve for Pt-5Mo vs Pt-13Rh in Figure 3-3 may be somewhat low for the reason given in discussing Pt-5Mo vs Pt; that is, that Pt-Mo voltage versus alloy content results of Nishimura showed a decreasing output with increasing molybdenum content in this percentage range, contrary to other results in the literature. Therefore his Pt-5Mo may have had an unusually low thermoelectric power relative to platinum or Pt-13Rh.

Pt-1Re vs Pt, Pt-2Re vs Pt.   These couples were calibrated from 0 to about 1350°C in a reducing atmosphere by Sims, Craighead, and others (1956), and the results are given in Figure 3-4. Earlier work was reported by Schulze (1933). The thermoelectric power of the 1 per cent rhenium couple is about 19 μV/°C in the 1000-1200°C range. The couple containing 2 per cent rhenium has a slightly higher sensitivity, about 20 μV/°C in the same range.

The addition of one per cent rhenium to platinum increases the tensile strength nearly fourfold, with a corresponding decrease in ductility. On the basis of 100 hours exposure to air at 800°C, followed by measurement of the weight loss, Sims, et al., suggested that the platinum-rhenium alloys are less resistant to oxidation than pure platinum.

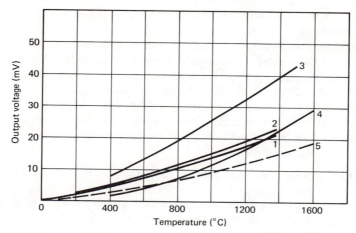

FIGURE 3-4   Thermocouples with rhenium or platinum-rhenium alloys versus platinum or rhodium.   1. Pt-1Re vs Pt, Sims, Craighead, et al. (1956).   2. Pt-2Re vs Pt, Sims, Craighead, et al. (1956). 3. Pt-8Re vs Pt, Zysk (1964).   4. Re vs Pt, Sims and Jaffee (1956). 5. Pt-8Re vs Rh, Schulze (1939).

Pt-8Re vs Pt. Schulze (1939) described work that included a calibration to about 1330°C; he stated that the platinum-rhenium alloy had good workability, and recommended it for that reason as well as for its resistance to oxidation effects. Recrystallization and subsequent brittleness after heating were reported, however. The effect was observed after 100 hours at 1300°C. Stauss (1941) later commented on the relatively high output, but stated that the couple was subject to a more rapid change of calibration than Pt-10Rh vs Pt. Zysk (1964) gave a calibration up to 1500°C but did not recommend use in air at temperatures above 600°C. Condensed results from the calibration are plotted in Figure 3-4. Using Zysk's data, the thermoelectric power is found to average about 35 μV/°C in the 1100 to 1500°C range.

Re vs Pt. A brief description of this thermocouple was given by Sims and Jaffee (1956), and Sims, Craighead, et al. (1956). The voltage-temperature curve from this work is included in Figure 3-4. Calibrations were made in an atmosphere of hydrogen as well as argon, and both sets of data were in excellent agreement. The thermoelectric power is 23.8 $\mu V/°C$ and 39.3 $\mu V/°C$ at 1000 and 1600°C, respectively, and is very low near room temperature.

Pt-8Re vs Rh.   No detailed information exists in recent literature, although a number of reviews mention this couple: Schulze (1939), Sanders (1958), Bennett (1961, p. 28), and Stadnyk and Samsonov (1964). The calibration curve is shown in Figure 3-4, from Schulze, and the thermoelectric power S is about 14 $\mu V/°C$ in the 1000-1100°C region.

Stadnyk and Samsonov considered this as a thermocouple for oxidizing atmospheres, but indicated a relatively high rate of drift during a ten-hour use period. This consisted of a 55°C change in calibration compared with a 1.5°C change for a Pt-10Rh vs Pt couple under similar circumstances.

The Pt-Re alloy has been described as subject to recrystallization with subsequent brittleness, and as a contributor to thermocouple instability at higher temperatures. The latter condition is due to oxidation of the rhenium, with formation of Re_2O_7, a volatile oxide.

Re-10Pt vs W-26Re.   An investigation by Kuhlman (1963, May; 1963, Sept.) led to rather unfavorable conclusions for this couple. The rhenium-platinum alloy was found to be unstable at room temperature in air, which was shown by a surface discoloration. However, there was no thermoelectric instability noted during measurement tests.

Te vs Pt.   Schulze (1939) reviewed special thermocouples with tellurium, used for the measurement of radiant heat. Te vs Pt was said to be useful at temperatures from −75 to +90°C. At higher temperatures rapid drift takes place. This, and the high resistivity of tellurium, are considerable disadvantages. A high thermoelectric power of about 400 $\mu V/°C$ was given, however.

V vs Pt.   Bliss (1965) investigated vanadium and other base elements as possible high temperature materials for nuclear applications. Druzhinina, et al. (1966) calibrated nine thermocouples over the range of 0 to 1600°C in vacuum. The graph shown in Figure 3-5 is a mean value curve for all calibrations. The thermoelectric power is 18.6 and 28.5 $\mu V/°C$ at 500 and 1000°C, respectively.

The vanadium wire had a fibrous structure in the cold-hardened condition, with high tensile strength and low elongation. After annealing for 30 minutes at 1000°C, the wire acquired a recrystallized structure with satisfactory strength and elongation. The wire could be easily wound into coils of diameter four to six times larger than the wire diameter, both in the cold drawn and the annealed conditions.

Vanadium was recommended as a thermocouple material for prolonged measurements at 1300°C and for brief periods up to 1500°C, presumably in vacuum because of the tendency of the vanadium to oxidize. The possibility of inert atmosphere use was not discussed, but there appear to be no reasons preventing such application.

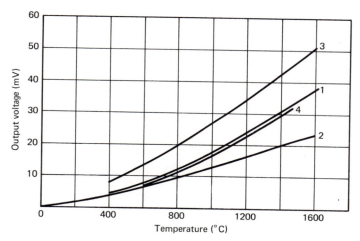

FIGURE 3-5   Thermocouples with noble elements plus vanadium or tungsten.   1. V vs Pt, Druzhinina, et al. (1966).   2. Pt-3W vs Pt, Nishimura (1958).   3. Pt-9W vs Pt, Zysk (1964).   4. Pt-10Rh vs W, Van Liempt (1929).

Pt-3W vs Pt.   This couple was among those cited in a patent by Nishimura (1958) claiming thermocouples with properties similar to the Pt-10Rh vs Pt couple, but with higher outputs. The calibration curve is included in Figure 3-5. The thermoelectric power is about 18 $\mu V/^{\circ}C$ from 900 to 1600°C.

Pt-9W vs Pt.   Zysk (1964) commented on data presented for this couple, stating that use in air is not recommended above 600°C. As shown in Figure 3-5, the output is considerably greater than that of Pt-3W vs Pt. The thermoelectric power is 41 $\mu V/^{\circ}C$ in the 1300-1600°C range.

Pt-10Rh vs W.   This was developed by Van Liempt (1929) and was one of the first modern-day thermocouples that paired noble and base metals for high temperature measurements. Van Liempt attempted to improve upon the performance of the Pt-10Rh vs Pt pair in reducing atmospheres by substitution of tungsten for the platinum. The calibration curve is given in Figure 3-5. The thermoelectric power is 34 $\mu V/^{\circ}C$ between 1200 and 1400°C. An upper temperature limit of 1830°C was recommended by Van Liempt.

Elemental and Binary Alloy Thermocouples Not Containing Platinum

Ag vs Al.   Thin film couples prepared by evaporative technique were used by Bailey, Richard, and Mitchell (1969) for measuring the temperatures of Permalloy films on glass substrates during repetitive temperature cycling between room and cryogenic temperatures. The couples were found to agree closely with a wire thermocouple of the same elements, when calibrated in the 4.2 to 300 K temperature interval. The voltage-temperature curve is linear from 27 to 300 K. The thermoelectric power is 3.1 μV/K for that interval of temperature. This represents a low sensitivity; however, the degree of linearity is unusual for these temperatures, which may be a compensating advantage.

The thin film thermocouples were reported to be stable and reproducible. At 77 K a repeatability of ±0.1 K was observed. The film and wire couples were in near agreement, varying at most by 1.7 per cent at 4.2 K.

Ag vs Bi.   According to Hornig and O'Keefe (1947), this combination has been used in thermopiles for radiometry. The thermoelectric power was given as about 63 μV/°C. Day, Gaddy, and Iverson (1968) used a thin film pair as a fast response detector for molecular lasers operating in the middle and far infrared. Crossed strips of silver and bismuth were vapor deposited on a substrate of beryllium oxide to thicknesses of 1000 Å. The films were noticeably grainy, due to the polycrystalline nature of the substrate; the film resistance was several times the computed resistance using bulk conductivity data.

Ag vs Ni.   A thin film couple using these elements was developed by Sandfort and Charlson (1968) for studies of the temperature variation across semiconductor thin films. The calibration curve was linear in the 20 to 50°C temperature range, with a thermoelectric power of 19 μV/°C. Measurable drift was observed after two weeks of continuous exposure at 55°C. A 100 Å thickness thermocouple drifted about –1.4 per cent. For a 1000 Å couple drift was near –2 per cent. Similar thermocouples of Au vs Ni were somewhat more stable in performance.

Ag vs Cu-43Ni.   Silver-constantan thermocouples can be used at somewhat higher temperatures than Cu vs Cu-43Ni because of the improved oxidation resistance of silver. A voltage-temperature curve from data tabulated by Roeser and Wensel (1941, p. 1293) is included in Figure 3-6. The thermoelectric power averages 66 μV/°C in the 400 to 600°C range. Schulze (1939) described use of this combination in Germany and recommended it for temperatures up to 600°C compared with a 500°C limit for copper-constantan.

Silver-constantan was used by Gier and Bolter (1941) in developing sensitive thermopiles for a radiometer, and by Hartwig (1957) for an aerodynamic heat flux sensor. Both of these applications utilized constantan thermocouple wire upon which a layer of silver was deposited in selected areas; the wire was mounted on a form that positioned the silver coated and bare constantan sections to form the thermopile. Pesko and associates (1971) analyzed the performance of plated thermocouples and developed an expression for the output voltage in terms of the plating cross sectional area and other parameters. The result is applicable to this type of thermopile, although the experimental portion of the work was done with Cu vs Cu-43Ni.

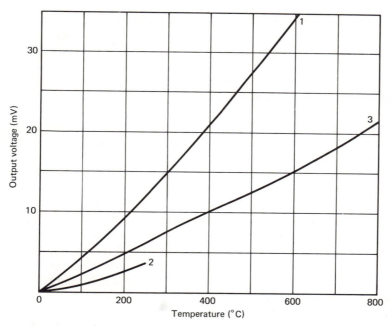

FIGURE 3-6   Silver and gold versus base metals. 1. Ag vs Cu-43Ni, Roeser and Wensel (1941, p. 1293).   2. Au vs Ni, King, et al. (1968).   3. Au vs Ni, Roeser and Wensel (1941, p. 1293).

Ag vs PbTe.[3]   Day and coworkers (1968) briefly investigated a thin film couple of this combination because of the high thermoelectric power, but found that the response time of thin film metal-semiconductor junctions was substantially slower than that of similar metallic thermocouples. No other details were given.

Te-33Ag vs Cu.   This was developed by Hatfield and Wilkens (1950) for use as a difference thermocouple in a high sensitivity heat flux sensor. The tellurium-silver alloy was free of the drift associated with commercially pure tellurium, while having an even larger value of thermoelectric power. There was no noticeable drift in the 15 to 100°C region and use without damage for at least short times to 200°C was indicated.

Au vs Al. A thermocouple of this type was used by Freeman and Bass (1970) during evaluation of an apparatus for determining thermoelectric power. Small difference temperatures were employed throughout a range of from 30 to 150°C.

Au vs Bi. Thin film couples on fused silicon dioxide substrates were used by Davies (1970) during studies of radiant heating of simulated human skin. The film thicknesses varied from 1500 to 6000 Å.

Au vs Ni. Thin film thermocouples were described in several reports: Sandfort and Charlson (1968), King, Camilli, and Findeis (1968), Moeller, Noland, and Rhodes (1968), and Thornburg and Wayman (1969). Quartz and glass were used as substrates. Also, in one instance, Moeller, et al., described the use of a nickel plate as both the substrate and one conductor. The gold film was deposited on a layer of silicon monoxide attached to the substrate, and overlapped onto the nickel in the area of the gold-nickel junction.

King, Camilli, and Findeis reported results of work at temperatures as
Sandfort and Charlson, who also investigated Ag vs Ni, preferred the gold-nickel pair. Both gold and silver make good ohmic content with a wide range of semiconductors, which was an advantage for their investigation of temperature variation across thin semiconductor films. The voltage-temperature characteristic was linear in the 20 to 50°C range with a thermoelectric power of 20 μV/°C. Output was independent of film thickness, at least for conditions where temperature gradients in the films leading from the junction area were small. No thermocouple drift was observed during several weeks use at temperatures up to 50°C. Detectable drift appeared at higher temperatures. The voltage output of a 100 Å thick couple decreased 5 μV while that of a 1000 Å unit decreased 3 μV during a two-week exposure at 55°C. The 5 μV drift represented a change of –0.45 per cent of the output at 55°C, using the data given by the authors.

King, Camilli, and Findeis reported results of work at temperatures as high as 400°C, in developing a thin film difference thermocouple for thermal analysis work. Figure 3-6 includes a calibration curve over the range of 0 to about 250°C. Thermoelectric power increased from 19 μV/°C at 100°C to 24.5 μV/°C at 200°C. Drift of 0.5 μV was reported during tests from room temperature to 400°C.

Thornburg and Wayman vapor-deposited gold and nickel on substrates of Corning 7059 glass, said to be capable of withstanding prolonged thermal treatment at 300°C; this glass also has excellent surface finish and exhibits a high electrical resistivity. All post-preparation work was done in a dry nitrogen atmosphere to prevent specimen oxidation. Best results were obtained when each film was annealed and outgassed by heating to 300°C for one hour in the high 10^{-7} torr range.

Unannealed couples lacked stability of thermoelectric power, showing a day-to-day negative drift that was caused by recrystallization. Well-annealed specimens, which were not outgassed, yielded time-stable calibration curves, but showed excessive scatter in a plot of the dependence of thermoelectric power on film thickness.

Voltage calibrations of annealed outgassed specimens were made from room temperature up to about 150°C. The data compared well with that of a wire couple of the same metals. An equation for the thermocouple voltage V was found, given by

$$V = 23.387T + 6.88 \times 10^{-3} T^2$$

where V is in microvolts and T is in degrees Celsius. The thermoelectric power of 23.4 $\mu V/°C$ compared favorably with a value of 22.5 $\mu V/°C$ obtained from measurements of a gold-nickel wire thermocouple. The thickness dependence of thermoelectric power was found to be slight, and did not markedly deviate from bulk values until the thickness of the nickel was decreased below about 500 Å. The thickness of the gold was about 1400 to 1500 Å, said to be in the bulk value range.

A curve for bulk metal Au vs Ni using earlier data from Roeser and Wensel (1941, p. 1293) is included in Figure 3-6. There is good agreement with the equation of Thornburg and Wayman.

Ag-0.37at.Au vs Au-2.1at.Co. This two-alloy pair was first suggested for cryogenic work by Borelius, et al. (1932), and Keesom and Matthijs (1935), before it was realized that the cobalt-bearing alloy was metallurgically unstable at slightly above room temperature. Fuschillo (1954) indicated that it is difficult to obtain Au-2.lat.Co wire in a homogeneous condition and described methods for checking homogeneity and selecting homogeneous wires of this and other alloys for cryogenic use. Zysk (1964) stated that homogeneity must be achieved before drawing the wire and great care must be taken to maintain the condition during the drawing process. Other work by Fuschillo (1957) described use of this alloy in a couple paired with copper.

Results by Powell, Bunch, and Caywood (1961) with Cu vs Au-2.lat.Co. are also applicable here, since the instability and drift problems are due to the gold-cobalt. In fact there is little difference in the voltage-temperature characteristics of the two couples, since copper and Ag-0.37at.Au are thermoelectrically similar, as was pointed out by Powell, Bunch, and Corruccini (1961); this is discussed in detail under Cu vs Ag-0.37at.Au. An important property of the silver-gold alloy is its low thermal conductivity, which is desirable in cryogenic work to avoid heat leaks. The alloy is well known by the name normal silver.

The Ag-0.37at.Au vs Au-2.lat.Co couple can be used in the 4 through 300 K temperature range, provided that no high temperature preparation steps such as welding of the junction are used. Short-time exposure to temperatures above room temperature may be sufficient to precipitate some of the cobalt, and there may be some room temperature aging, or drift, with time at that temperature. According to Powell, Bunch, and Caywood (1961) a 24-hour "anneal" at 90°C caused a 1 per cent decrease in thermoelectric power.

Ag-0.37at.Au vs Au-(0.02-0.03)at.Fe.   The large thermoelectric power of gold-iron in the cryogenic region was discovered by Borelius and coworkers in 1932. However, the properties of gold-iron were not fully appreciated until the work by MacDonald, Pearson, and Templeton (1962). When gold-iron was paired with normal silver, Berman, Brock, and Huntley (1965) found the useful temperature range to be about 1 to 40 K. Voltage-temperature curves and other data by Berman and Huntley (1963) indicate a maximum thermoelectric power S of about 13.6 μV/K near 10 K for the 0.03at.Fe alloy, and at a slightly lower temperature for the 0.02at.Fe alloy. Depending on the precise amount of iron, there can be some variations in these figures. Schriempf and Schindler (1965, 1966) gave a maximum value of 14 μV/K at 8 K, and 10 μV/K at as low as 1 K in temperature. These investigators demonstrated that some earlier reports of anomalies in the voltage-temperature curve were due to measurement difficulties and not to the characteristics of the thermocouple. From Berman, Brock, and Huntley (1965), different samples of gold-iron wire from the same batch may vary up to 1.5 per cent in the value of S. Above 40 K, S is low. A voltage-temperature curve from Sparks and Hall (1969) is shown in Figures 3-7 and 3-8 for Ag-0.37at.Au vs Au-0.02at.Fe.

The wires have been used both as-received and in the annealed condition. Kutzner (1968) used wires as received from the manufacturer, and the repeatability was better than 0.5 per cent for an application where the wires did not pass through any region hotter than the warmer junction. Liquid helium or nitrogen was used at the latter. Berman and Huntley (1963) annealed all specimens in vacuum at 550°C for at least 12 hours after drawing the wires. A repeatability of 0.2 per cent above 4 K was obtained.

The thermocouple wires should be handled carefully to avoid excessive strain. Such straining of the gold-iron wire may produce changes in thermoelectric power of as much as 5 per cent; normal handling during installation contributes negligible effects. According to Berman, Brock, and Huntley (1964), calibration drift was less than 0.5 per cent during a two-year period of use for a number of thermocouples tested.

Berman, et al., also determined the change of thermoelectric power of normal silver vs gold-iron in a transverse magnetic field. A change of as much as 20 per cent can occur at about 1 K in a field of 20 kG. It was stated, however, that such a change can be accurately calibrated, so that corrections can be applied to temperature measurements. Berman, et al. (1968), and Huntley and Walker (1969) have discussed theory and observations of voltage dependence on magnetic field for Au-0.03at.Fe and other dilute gold-iron alloys.

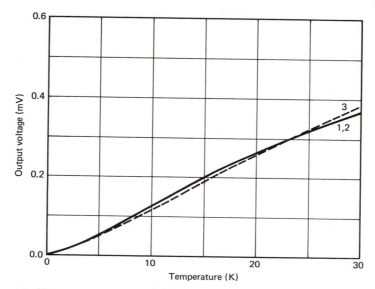

FIGURE 3-7 Cryogenic thermocouples, I. Range of 0 to 30 K is shown. The reference junction temperature is at 0 K. 1. Ag-0.37at. Au vs Au-0.02at.Fe; 2. Cu vs Au-0.02at.Fe; 3. Cu vs Au-0.07at. Fe: Sparks and Hall (1969).

Ag-0.37at.Au vs Cu-43Ni. Corruccini (1963) described this as a somewhat improved version of the copper-constantan couple for cryogenic use. By replacing copper with normal silver, the disadvantage of the high thermal conductivity of copper is eliminated. Also, normal silver is less thermoelectrically sensitive to impurities than copper, at and below the temperature of liquid hydrogen. Details on the silver-bearing alloy are given under Cu vs Ag-0.37at.Au.

Ag-28at.Au vs Cu-43Ni. The substitution of Ag-28at.Au for copper in copper-constantan was discussed by Corruccini (1963) for cryogenic thermocouples. This silver-gold alloy has an even lower thermal conductivity than normal silver, and is also less sensitive to variations in composition.

Cu vs Ag-0.37at.Au. A comparison of copper and the alloy normal silver is given here. The two metals are thermoelectrically similar and are not used together as a thermocouple; one is sometimes substituted for the other, however, in a couple such as copper-constantan.

The alloy Ag-0.37at.Au was developed at the Kamerlingh Onnes Laboratory of Leiden University during the 1920-1930 decade by Borelius (1953) and others, and was used as a laboratory standard or "normal." It is now often referred to as normal silver (or as silver normal). It has the advantage of a low thermal conductivity, but is susceptible to cold work, is sensitive to the amount of gold added during manufacture, and is also affected by impurities. Other alloys such as Ag-4.8at.Ga, described by Crisp and Henry (1964), and Ag-28at.Au, by Corruccini (1963), and Sparks and Powell (1967), have been suggested more recently as replacements, but the use of normal silver for cryogenic work is widespread.

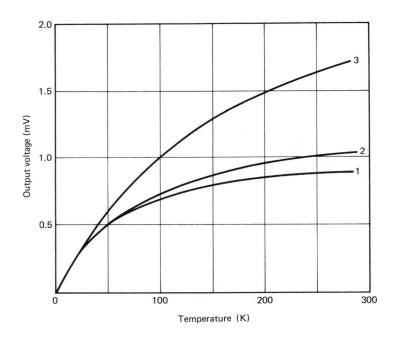

FIGURE 3-8   Cryogenic thermocouples, II. Range of 0 to 300 K is shown. The reference junction temperature is at 0 K.   1. Ag-0.37 at.Au vs Au-0.02at.Fe;   2. Cu vs Au-0.02at.Fe;   3. Cu vs Au-0.07 at.Fe: Sparks and Hall (1969).

Both copper and normal silver have approximately the same thermoelectric power in the 0-300 K range. A comparison of the thermoelectric voltage and power is shown in terms of a copper vs normal silver thermocouple in Table 3-1, from Powell, Bunch, and Corruccini (1961), and Powell, Caywood, and Bunch

(1962). The cold junction is referenced by computation to 0 K so that maximum output exists with the warm junction at room temperature. Thermocouple grade copper is slightly more homogeneous than normal silver. On the other hand, the alloy was said by Corruccini (1963) to be less thermoelectrically sensitive to impurities, for temperatures near that of liquid hydrogen. In general, copper is preferred above 80 K, while normal silver becomes preferable below 80 K where heat leaks along wires can be important.

TABLE 3-1

Voltage and Thermoelectric Power of a Copper vs Normal Silver Thermocouple with the Cold Junction at 0 K

Temperature (K)	Voltage (μV)	Thermoelectric Power (μV/K)
0	0.00	0.00
5	0.00	0.00
20	0.2	0.06
50	15.2	0.86
100	57.2	0.73
200	121.8	0.63
300	192.3	0.77

Cu vs Au-2.1at.Co.   Copper versus gold-cobalt is very similar to Ag-0.37at.Au vs Au-2.lat.Co, which was discussed previously.   Fuschillo (1957) reviewed the work by Borelius and others that proposed the use of Au-2.lat.Co for a cryogenic thermocouple, and developed a low temperature reference scale using copper with gold-cobalt rather than the normal silver suggested by Borelius.   This was reported to decrease uncertainties in the use of the resulting temperature reference table.   Because of earlier experience with homogeneity studies, Fuschillo recommended testing the gold-cobalt wire for homogeneity prior to use and also recommended minimizing wire temperature gradients as much as possible. The possible existence of gold-cobalt instability was mentioned.

Other work during this period included the use of eight couples in a cryogenic thermal conductivity apparatus for evaluating copper specimens, reported by Powell, Roder, and Rogers (1957). Scientists at the National Bureau of Standards also calibrated copper vs gold-cobalt. Data by Powell, Caywood, and Bunch (1962) are plotted in Figures 3-9 and 3-10. The information is based on a cold junction reference computed for 0 K. The thermoelectric power was 4.04 μV/K at 4 K and 16.43 μV/K at 20 K.

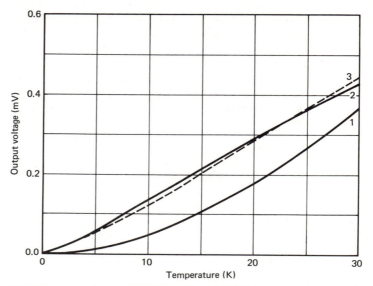

FIGURE 3-9   Cryogenic thermocouples, III. Range of 0 to 30 K is shown. The reference junction temperature is at 0 K.   1. Cu vs Au-2.1at.Co, Powell, Caywood, and Bunch (1962).   2. Ni-10Cr vs Au-0.02at.Fe, Sparks and Hall (1969).   3. Ni-10Cr vs Au-0.07at.Fe, Sparks, Powell, and Hall (1968).

Problems with instability led to 2.5-year drift tests by Powell, Bunch, and Caywood (1961) to determine whether room temperature aging was significant. Of three thermocouples tested, one showed no significant decrease in thermoelectric power, one decreased 3 per cent, and one decreased 5 per cent. For the latter two, approximately a 1 per cent change occurred in the first six months. A short-time test consisting of a 24-hour "anneal" at 90°C caused a 1 per cent decrease in thermoelectric power.

This thermocouple can be used in the 4 through 300 K range, provided that no high temperature preparation steps such as welding of the junction are used. It is insensitive to magnetic fields when the section within the field is at constant temperature. This was shown by Richards, Edwards, and Legvold (1969) for fields up to 100 kG in liquid helium.

Cu vs Au-(0.02-0.03)at.Fe, Cu vs Au-0.07at.Fe.   These couples are similar to the normal silver vs gold-iron cryogenic combinations discussed earlier. Several groups of investigators have worked with the gold-iron alloys: Berman and Huntley (1963), Berman, Brock, and Huntley (1964), Rosenbaum, Oder, and Goldner (1964), Schriempf and Schindler (1965, 1966), and Medvedeva, et al. (1971, p. 316). Voltage-temperature curves from results by Sparks and Hall (1969) appear in Figures 3-7 and 3-8 for Cu vs Au-0.02at.Fe. From the same work, the thermoelectric power S is 9.15 μV/K at 1 K, a maximum of 14.58 μV/K at 9 K, and 11.95 μV/K at 20 K. Similar results were found for Cu vs Au-0.07at.Fe, Figures 3-7 and 3-8. In this case, S is 8.48 μV/K at 1 K, a maximum of 14.15 μV/K at 12 K, and 13.38 μV/K at 20 K.

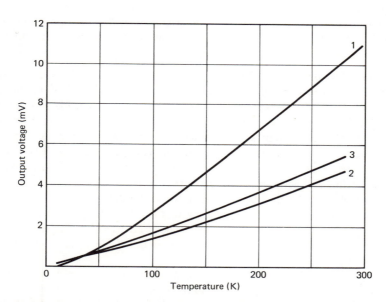

FIGURE 3-10   Cryogenic thermocouples, IV. Range of 0 to 300 K is shown. The reference junction temperature is at 0 K.   1. Cu vs Au-2.1at.Co, Powell, Caywood, and Bunch (1962).   2. Ni-10Cr vs Au-0.02at.Fe, Sparks and Hall (1969).   3. Ni-10Cr vs Au-0.07at. Fe, Sparks, Powell, and Hall (1968).

Richards, Edwards, and Legvold (1969) experimentally verified their analytical work on effects of magnetic fields upon thermoelectric power by showing that copper vs gold-iron is insensitive to fields up to 100 kG, provided that the segment in the field is at constant temperature. This was shown for temperatures from the liquid helium region to 300 K.

The couple utilizing Au-0.07at.Fe was also investigated by Finnemore, Ostenson, and Stromberg (1965). A thermoelectric power of 13.7 ±0.3 μV/K in the 6 to 20 K range was reported. At 78 K the thermoelectric power was 8 μV/K. A repeatability of better than 0.02 per cent on successive cool-downs and an accuracy of 0.02 K in the range from 4 to 20 K were also shown.

Dworschak and coworkers (1970) used the Cu vs Au-0.03at.Fe couple for temperature measurements of metallic resistivity specimens exposed to electron irradiation from a particle accelerator. Temperatures were in the liquid helium range. The wire diameter was 0.08 mm and the conductors were soldered to the specimen at the measuring junction.

Ag-0.37at.Au vs Cu-(0.1-0.3)at.Fe.   Fuschillo (1954) mentioned this combination and indicated that it had been among those couples suggested by Borelius and others in the early 1930's for use at temperatures down to 2.5 K. The thermoelectric power was reported to be several times that of copper-constantan at 20 K. It was also suggested that the homogeneity of the copper-iron wire might prove to be better than that of Au-2.lat.Co.

Nb vs Au-(0.02-0.03)at.Fe.   Niobium was used by Berman, Brock, and Huntley (1964, 1965) in place of normal silver below about 9 K where niobium becomes superconductive. Heat leakage could then be minimized by using very fine wire without paying the penalty of increased wire resistance. Further, the thermoelectric power of the niobium superconductive section was zero, as is true for all superconductors in that state. Thus spurious voltages could not occur in the superconductive section. Calibration drift was 0.5 per cent or less after repeated thermal cycling and periods of time up to two years.

Rosenbaum (1970) suggested that Nb vs Au-0.02at.Fe may be usable to at least as low as 0.05 K, recommending that the diameter of the gold-iron wire be as small as possible to minimize heat leakage. It was also suggested that the reference temperature be maintained at about 0.3 K, in order to be near the measuring junction temperature.

Berman and Rogers (1964) reported a thermoelectric power of over 10 μV/K below 4 K, for a couple with Au-0.03at.Fe, and used this pair in an investigation of the thermal conductivity of solid helium in the 1.4 to 1.8 K region. Differences of temperature of about 0.01 K along the specimen were measured with an accuracy of 1 per cent. When necessary, differences as low as 4 millidegrees K were used.

Ni-10Cr vs Au-2.1at.Co.   Slack (1961) discussed Chromel[4] vs gold-cobalt[5] and other couples for use in cryogenic thermal conductivity measurements. Corruccini (1963) also discussed this couple, and Koeppe (1967) computed a voltage-temperature characteristic based on experimental data from calibration of each thermoelement against copper.

Chromel vs gold-cobalt was reported to have a relatively high thermoelectric power over the entire 3 to 300 K temperature range; values of 10.7 μV/K at 10 K and 64.0 μV/K at 273.2 K were given by Slack.[5] Another advantage for cryogenic work is the low thermal conductivity of both alloys, a property that can be used to minimize heat leakage. However, as was previously stated, the gold-cobalt alloy is metallurgically unstable above room temperature. Most of the discussion of this problem that was given for Cu vs Au-2.lat.Co is also applicable here, since copper and Chromel are both excellent thermoelements and are not contributors to the instability problem.

Ni-10Cr vs Au-(0.02-0.03)at.Fe. Work by Berman, Brock, and Huntley (1965) indicated a satisfactory thermoelectric power throughout the 1 to 300 K range. Therefore, a single couple can be used for the entire temperature interval between liquid helium and room temperatures. The gold-iron is the primary contributor to the thermoelectric power of the pair below about 40 K and the thermoelectric power of the Chromel-P becomes dominant at higher temperatures; this leads to a relatively smooth over-all curve of voltage vs temperature. A plot for Ni-10Cr vs Au-0.02at.Fe, using data from Sparks and Hall (1969), is shown in Figures 3-9 and 3-10. The thermoelectric power S is 9.32 μV/K at 1 K, reaches a maximum of 16.18 μV/K at 11 K, and then goes through a minimum of 12.39 μV/K at 42 K. At 280 K, S is 20.90 μV/K.

Calibration drift due to repeated thermal cycling and aging was reported by Berman, Brock, and Huntley (1965) to be equal to or less than 0.5 per cent. Applications for this couple have been reported by Ashworth and Steeple (1964), and Rechowicz, Ashworth, and Steeple (1969).

Calibration and use of the gold-iron wire in both hard drawn and annealed conditions has been described. Kutzner (1968) calibrated as-received Au-0.02at.Fe wire and showed a curve for the region between 4.2 and 170 K. With liquid helium or liquid nitrogen as the reference junction, the repeatability was better than 0.5 per cent; in this case the thermocouple wires did not pass through a region warmer than the warmest junction. Rosenbaum (1968, 1969) worked with both hard drawn and annealed Au-0.02at.Fe, and found that the annealing process reduced random voltage differences between sample couples. The annealing process also led to an increased thermoelectric power. The improvement produced by annealing resulted in an extension of earlier calibrations down to 0.55 K and allowed determination of thermoelectric power below 3 K to an accuracy of ±2 per cent. The accuracy was ±1 per cent for the equivalent data at higher temperatures, up to 85.2 K.

According to Rosenbaum, Chromel-P vs gold-iron was used at the University of Illinois for several years prior to 1968. During that period, it was found that annealed gold-iron wires from the same bar can be used with Chromel-P as absolute thermometers to as low as 0.4 K. However, each wire from the same bar must be calibrated against a known standard if accuracies of better than ±0.05 K are desired at temperatures below 1 K. Differences were found in the thermoelectric voltages of wires made from two different bars from the same manufacturer. A difference of 11.3 per cent was observed between two wires when the junctions were at 4.2 and 77 K. Therefore, bar-to-bar uniformity had not reached the point where standardized thermocouple tables could be used.

Rosenbaum also investigated the effects of work hardening on annealed gold-iron wires and found that normal handling produced a 0.25 per cent decrease in voltage compared with a wire that was very carefully handled, so that a thermocouple should be treated very gently when intended for use in the annealed condition. This pair appears to be somewhat less useful than Chromel-P vs Au-0.07at.Fe, which was also investigated, because of this property. Repeatability of the thermocouple was found to be excellent, however, and spurious thermoelectric voltages were negligible for determining liquid helium temperatures with the reference junction either in liquid nitrogen or at the ice point. Spurious voltages never exceeded 4 μV. Vacuum jacket encapsulation of the wires along the section in the 77-300 K temperature gradient proved to be unnecessary, although long lengths of wire of one-third to two-thirds of a meter were recommended to traverse that temperature interval.

Ni-10Cr vs Au-0.07at.Fe. A relatively high and uniform thermoelectric power led Sparks and Powell (1967), and Sparks, Powell, and Hall (1968, 1970) to suggest the use of this couple for the entire cryogenic range. It was also stated that this thermocouple is the most sensitive combination available for measurements below 20 K, and finally, that its usefulness is enhanced by the relatively linear voltage-temperature characteristic. A plot of voltage versus temperature is shown in Figures 3-9 and 3-10. It was reported that this curve has a maximum voltage deviation from linearity of about 5 per cent of full scale, which is an unusual degree of linearity for that temperature range. The thermoelectric power is 8.65, 15.56, and 16.31 μV/K at 1, 10, and 20 K, respectively. Rosenbaum (1968, 1969) also gave calibration tables for this couple, based on a 4.2 K cold junction temperature.

Rosenbaum showed that annealing of the Au-0.07at.Fe makes some difference in performance at lower temperatures. Whereas an annealed wire thermocouple showed a peak of 17 μV/K in thermoelectric power S at about 17 K, and a minimum of 15.7 μV/K at 40 K, a hard worked specimen showed no large maximum in the 10 to 40 K range; instead, there were three minor peaks with the value of S remaining at about 15 μV/K throughout that temperature interval.

Effects of handling the annealed specimens were also investigated, and it was observed that normal handling did not decrease the thermoelectric voltage, as was found in the case of the more dilute Au-0.02at.Fe alloy. However, a decrease of 0.42 per cent in voltage was observed with harshly handled

Au-0.07at.Fe; a value of –1.42 per cent resulted for the more dilute alloy. The largest deviations occurred below 20 K. It was concluded that this couple can be used in the annealed condition with confidence of not hard working the wire under normal handling. However, if rough handling is necessary, it was suggested that hard drawn wire be used.

According to Rosenbaum, an advantage of using annealed wire lies in its improved properties in the range below about 3 K; variations in the voltages among annealed samples of wire are negligible down to 0.55 K. Thermoelectric power can be determined within ±2 per cent below 3 K with such specimens. Estimates of thermocouple calibration accuracy by Sparks, Powell, and Hall (1970) resulted in figures of ±0.010 K from 4 to 20 K, ±0.012 K from 20 to 75 K, and ±0.015 K from 75 to 280 K for the samples tested.

Sparks and Powell (1971) in their conference paper described the spread of output measurements from specimen to specimen of gold-iron alloy. This quantity, measured at 20 K for several specimens, was about 9 per cent of the total output at 20 K relative to a 4 K reference junction temperature. It was pointed out that spot calibrations are needed below 20 K because of this large deviation; also, this figure is at least indicative of the batch-to-batch wire uniformity that can be expected.

Von Middendorff (1971) evaluated the effects of strong magnetic fields at low temperatures and reported that for temperatures above 20 K S decreases, the greatest such decrease being about –2.5 per cent in the 35 to 60 K region, for a field of 60 kG. Below 20 K S increases, and a 40 per cent change results at 4.1 K, again with a field of 60 kG. No details on the field orientation or location relative to temperature gradients were given.

Strongin and associates (1971) chose this type of thermocouple for use in a liquid helium cryostat because it could withstand the high temperatures which were also required for cleaning their apparatus. They reported that Ni-10Cr vs Au-0.07at.Fe can be heated to about 1000°C without damage resulting.

Ni-10Cr vs Au-(9-10)Ni. Hicks and Valdsaar (1963) and Hicks (1969) discussed Au-9Ni for possible thermocouple applications. Dike (1954, p. 18) and Rosebury (1965, p. 531) mentioned a thermocouple of Ni-10Cr vs Au-10Ni which has a higher output than Chromel-constantan. Dike referred to the couple as Chromel vs white gold. The later work by Hicks, however, stated that extended exposure of Au-9Ni in the 400 to 600°C region revealed some phase separation. The metal was stable above and below that interval, and no instability was observed during short periods of time in the 400-600°C range. Greater amounts of nickel in the alloy lead to a wider range of instability. Smaller amounts result in lowered output.

Mo vs Ir. A calibration curve given by Troy and Steven (1950) is shown in Figure 3-11. The thermoelectric power, computed from the curve, is about 19

μV/$^\circ$C in the 1400-1800°C region. The calibration work was conducted in an atmosphere of helium. Sanders (1958) estimated an upper calibration limit and error of about 1930°C and $\pm$40°C, respectively. No details have been published on performance.

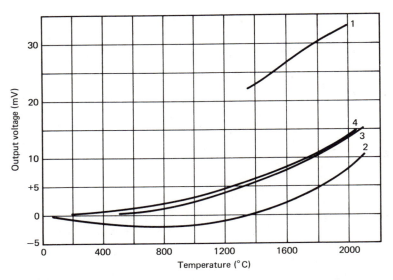

FIGURE 3-11   Thermocouples with iridium and other elements.
1. Mo vs Ir, Troy and Steven (1950).   2. Re vs Ir;   3. Re-30Ir vs Ir;   4. Re-40Ir vs Ir: Haase and Schneider (1956).

Re vs Ir.   Haase and Schneider (1956) included this pair in a study of the rhenium-iridium alloy system. The calibration curve is given in Figure 3-11. A calibration was also shown, together with other results, by Nadler and Kempter (1961). The output is negative at low temperatures and goes through a minimum near 850°C as shown in Figure 3-11. Above 1350°C rhenium is positive.

Re-30Ir vs Ir.   A voltage-temperature curve from Haase and Schneider (1956) is included in Figure 3-11. These investigators recommended the thermocouple because of its relatively large thermoelectric power of 14 μV/$^\circ$C between 1700 and 2000°C, and also because of a lack of sensitivity to small alloy composition changes. The latter property was said to permit usage in slightly oxidizing atmospheres as well as in vacuum or inert gases. A reduction of 10 to 15 per cent in the rhenium content of the alloy due to oxidation of the rhenium would result in a relatively small output drift,[6] amounting to about 30°C at 2000°C. The thermoelectric power is very low throughout the 0 to 500°C range, making cold junction compensation unecessary for many situations.

Re-40Ir vs Ir.   A curve from Haase and Schneider (1956) is given in Figure 3-11, taken from their results for the iridium-rhenium alloy system. The alloy is a less desirable choice than Re-30Ir if operation in a slightly oxidizing atmosphere is intended. Haase and Schneider stated that earlier mention of this combination in several articles was based on a misprinted designation for Rh-40Ir vs Ir in a paper which discussed the Feussner thermocouple. The latter couple is quite linear in output compared with Re-40Ir vs Ir, and its curve can be easily distinguished by that property.

Ir-10Rh vs Ir-10Re.   This couple was mentioned by Schulze (1939); use in oxidizing atmospheres is not recommended because the rhenium-bearing alloy tends to oxidize.

Re-30Ir vs Ir-20Re.   Information on this combination was given by Sanders (1958), based on a communication from P.E. Golsan. It was suggested for vacuum or inert atmosphere use. An output of 11.7 mV was reported at the maximum calibration temperature of about 2300°C. A calibration curve is shown in Figure 3-12, taken from data by Haase and Schneider (1956). Sanders estimated error for general use as ±45°C at 2300°C.

W vs Ir.   Troy and Steven (1950) made a detailed investigation of tungsten vs iridium during a program to develop a thermocouple with usefulness beyond the range of Pt-10Rh vs Pt. A calibration curve is included in Figure 3-12. Calibration accuracy in the 1100-2000°C interval ranged from 26 to 38°C when an optical pyrometer was used. The calibration uniformity, using wires from different lots in both the hard drawn and heat treated conditions, was within the calibration accuracy. Final calibrations were made by the melting point method using standard metals, and calibration error ranged from ±5 to ±5.8°C within the 1063-1966°C range. The thermoelectric power S varied from 22.9 μV/°C at 1000°C to 26.3 μV/°C at 2000°C. S was very low below about 50°C; this allows the elimination of cold junction compensation while still using calibration data based on an ice bath cold junction temperature. It was found that this procedure introduced an error of less than 0.1 per cent for hot junction temperatures above 1000°C.

Later articles and reviews, some with additional information, were given by several sources: Steven (1956), Sanders (1958), Kostkowski (1960), and Caldwell (1962). Generally, an upper temperature limit of 2000 to 2200°C was recommended for inert atmosphere or vacuum applications. According to Kostkowski, a 10°C calibration change may occur after ten hours at 2100°C in an inert atmosphere.

Troy and Steven investigated drift by means of a heating test which lasted for 120 hours at temperatures of 1900 to 2000°C in a helium atmosphere. No calibration change was observed for a thermocouple exposed to these conditions.

However, when a couple was exposed to a hydrogen atmosphere at these temperatures, progressive embrittlement and eventual breakage of the iridium wire occurred.

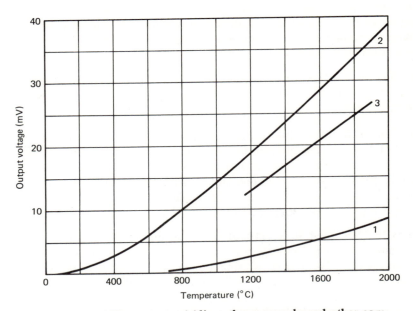

FIGURE 3-12   The tungsten-iridium thermocouple and other combinations with iridium alloys.   1. Re-30Ir vs Ir-20Re, Haase and Schneider (1956).   2. W vs Ir, Troy and Steven (1950).   3. W vs Rh-40Ir, Troy and Steven (1950).

Recrystallization also causes increased brittleness in both tungsten and iridium, and occurs during the first high temperature heating of the wires, regardless of the atmosphere. This is a disadvantage for the W vs Ir thermocouple. When the wires can be installed, calibrated in place, and left untouched during use, this property is not so important. Otherwise, the embrittled wires must be handled very carefully.

Pletenetskii and Mandrich (1965) conducted heat treatment tests with separate tungsten and iridium wires, and determined that tungsten heated in argon showed impaired stability during later heating in helium. Heat treatment to prepare for other tests consisted of heating tungsten for three minutes at 1800°C in vacuum or hydrogen. Iridium was heated for three minutes at 1800°C in vacuum, or in an oxidizing atmosphere for five minutes at 1700°C. These heat treated wires were then calibrated and heated at 1500°C in an atmosphere of helium with 0.1 per cent hydrogen. The results quoted were apparently typical for wires exposed to any of the initial heat treatments described (except for the argon atmosphere treatment of tungsten). A tungsten-iridium thermocouple showed a drift of $-1.2°C$ after the first five to ten hours. A slight continuing downward trend occurred during 50 hours of testing. The separate voltage curves of each wire versus an unspecified common reference material indicated that a net leveling off of the combined drift for W vs Ir may have been underway, at least during the last 20 hours. This was indicated by a continued downward drift of iridium, but with a reversal of the downward drift of the tungsten at 40 hours.

W vs Rh-40Ir. The calibration curve by Troy and Steven (1950) is shown in Figure 3-12, and was made during their search for a couple superior to Pt-10Rh vs Pt at high temperatures. From this curve, which was stated to be linear, the thermoelectric power can be computed to be about 19 $\mu V/°C$ between 1200 and 1900°C. The calibration was made in a helium atmosphere and the thermocouple voltage was 28.2 mV at 2000°C. Bennett, Hemphill, and Rainey (1966) conducted a drift test of Rh-40Ir with tungsten and other metals in an atmosphere of helium plus 100 ppm of CO at 1750°C. No appreciable drift was observed during a 90-hour exposure period.

Ni-10Cr vs Pd. Ihnat (1959), Clark and Hagel (1960), and Ihnat and Hagel (1960) investigated Chromel-palladium during a program to develop an improved jet engine turbine inlet thermocouple. The thermocouple arrangement consisted of a Chromel wire, protected by insulation and enclosed in a concentric outer tubing of palladium. The calibration curve is shown in Figure 3-13, from which the thermoelectric power appears to be above 50 $\mu V/°C$ in the 800-1000°C range.

Drift tests were conducted in a combustion chamber using fuel oil with 0.4 per cent sulphur to simulate the atmosphere in a jet engine. The exposure of a couple to these conditions over a 60-hour period at 1260°C resulted in output drift that reached a deviation of -0.7 per cent from the original output, after some 50 hours had passed. Failure of the couple occurred at this time, and it was found that the Chromel wire had oxidized in spite of the presumed protection by the insulation and palladium tubing.

Mo vs Pd, Ni-18Mo vs Pd, Ti vs Pd, W vs Pd. Ihnat (1959) presented calibration curves, which are given in Figure 3-13. The thermoelectric powers of all of these couples are relatively high. Ihnat evaluated the Ni-18Mo vs Pd, Ti vs Pd, and W vs Pd couples in the same manner as the Chromel-palladium couple that was described earlier. Time-to-failure figures were 50, 25, and 18 hours, respectively, for these three combinations. In each case, it was the base metal that had oxidized to failure. These thermocouples were of insulated concentric

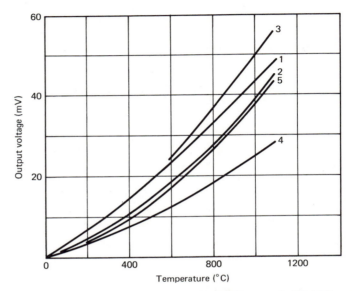

**FIGURE 3-13   Thermocouples with palladium.   1. Ni-10Cr vs Pd;
2. Mo vs Pd;   3. Ni-18Mo vs Pd;   4. Ti vs Pd;   5. W vs Pd: Ihnat
(1959).**

W vs Rh.   This pair was studied by Troy and Steven (1950) in their thermocouple investigation. The calibration curve, shown in Figure 3-14, ranged from
about 1130 to 2000°C. A helium atmosphere was used. The thermoelectric
power can be found from this curve to be about 23 μV/°C in the region of
1700°C. Additional data from Roeser and Wensel (1941, p. 1293) are plotted in
Figure 3-14 to show the voltage output at lower temperatures. The lack of
agreement between the two calibrations in the 1100°C region is unusually large.

Rh-8Re vs Rh.   Schulze (1939) presented a curve for this couple which is included in Figure 3-14. The thermoelectric power is slightly less than 6 μV/°C
in the 1600-1900°C interval. Earlier, Goedicke (1931) had recommended its use
up to 1900°C, and, in comparison with Pt-8Re vs Rh, found it the more stable of
the two.

Noble-Base Binary Alloys vs W-Re Alloys

All of the work reported in this section was done by Kuhlman (1963, May;
1963, Sept.; 1966) except for some of the information on W-0.5at.Os vs W-25Re,

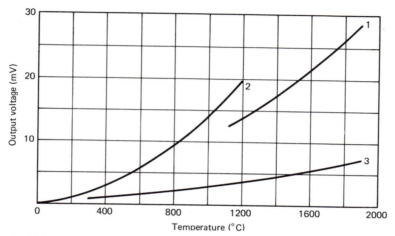

FIGURE 3-14 Thermocouples with rhodium. 1. W vs Rh, Troy and Steven (1950). 2. W vs Rh, Roeser and Wensel (1941, p. 1293). 3. Rh-8Re vs Rh, Schulze (1939).

which was a contribution by Funston and Kuhlman (1966). In much of the work, the objective was a suitable thermocouple for use up to 2500°C in a glide-reentry vehicle. Some of the tungsten-osmium alloy investigations were for a nuclear application.

The specimens used by Kuhlman were in the form of rods about 30 cm long by 0.32 cm in diameter, because wires of most of the required refractory metals would have been very difficult to draw. The rods were prepared by powder metallurgy techniques and were evaluated thermoelectrically after a final high temperature sintering process. In some cases the final rod composition differed considerably from the original powder composition; the latter is referred to here as the nominal composition. In such cases, both the nominal and final compositions are stated. Samples found to be two-phase rather than homogeneous after sintering are not included here. The W-26Re conductor in Kuhlman's work was used as a reference material and was not necessarily intended as the second metal in a final thermocouple selection. An alloy of Ir-20Re was suggested[7] for use as the negative conductor in some cases, as an alternative to W-26Re. Nonoxidizing atmospheres were used for the preparation work; calibrations were made in a helium atmosphere.

Re-10Os vs W-26Re, Re-20Os vs W-26Re.   Calibration curves are given in Figure 3-15 from Kuhlman's tabulated data. The osmium alloys were not as strong as tungsten, but nearly as brittle, and were not very different thermoelectrically from pure rhenium.

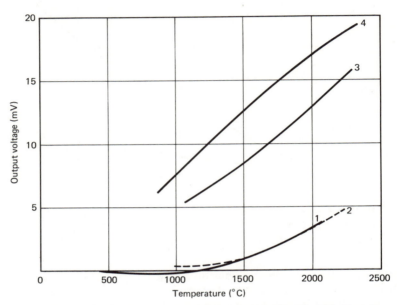

**FIGURE 3-15   Noble-base metal alloys paired with W-26Re, I.
1. Re-10Os vs W-36Re;   2. Re-20Os vs W-26Re;   3. Re-7.7Rh vs
W-26Re;   4. Re-15Rh vs W-26Re: Kuhlman (1963, May).**

Re-10Rh vs W-26Re, Re-20Rh vs W-26Re (Nominal Compositions).   The final compositions of the rhodium alloys were Re-7.7Rh and Re-15Rh, because of weight loss of rhodium during sintering of the bars. These alloys were believed to be promising for short time use at fairly high temperatures, but not as high as the 2500°C range where rhodium weight loss was observed. Calibration curves are shown in Figure 3-15. The thermoelectric power of Re-10Rh vs W-26Re ranged from 7.4 to 9.9 μV/°C, approximately, in the 1200-2300°C range. The higher rhodium content pair had values of about 9.9 to 5.5 μV/°C over the same interval, showing a reversed trend with temperature.

Re-5Ru vs W-26Re, Re-10Ru vs W-26Re, Re-15Ru vs W-26Re.   There was little or no weight loss observed by Kuhlman during preparation of the ruthenium alloys. The calibration curves for the 5 and 10 per cent pairs are given in Figure 3-16. The thermoelectric powers of the 5 and 10 per cent alloys changed somewhat with temperature. At 2300°C the 5 per cent ruthenium couple had a value near 8.1 $\mu V/°C$; that of the 10 per cent ruthenium couple was about 9.6 $\mu V/°C$. The Re-15Ru vs W-26Re couple was found to have a thermoelectric power near 11 $\mu V/°C$ at 2200°C. These alloys were considered to be among the best of those tested for application above 2500°C. The 5 per cent ruthenium alloy was not as strong as tungsten but was much more ductile.

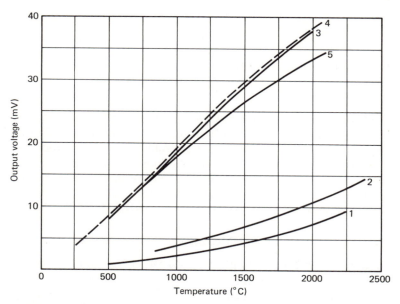

FIGURE 3-16   Noble-base metal alloys paired with W-26Re, II. 1. Re-5Ru vs W-26Re;   2. Re-10Ru vs W-26Re: Kuhlman (1963, May).   3. W-0.5at.Os vs W-26Re;   4. W-1at.Os vs W-26Re;   5. W-3at.Os vs W-26Re: Kuhlman (1966).

W-0.5at.Os vs W-(25-26)Re.   Kuhlman (1965, 1966) suggested the use of tungsten-osmium alloys for nuclear applications; the alloy with osmium was employed to decrease the sensitivity of the W vs W-26Re thermocouple to transmutation. Higher and more nearly linear output also resulted. A voltage-temperature curve is shown in Figure 3-16. Simulated transmutation of this thermocouple was described by Funston and Kuhlman (1966) and consisted of determining the outputs of couples with compositions equivalent to those of transmuted thermocouples after 0, 0.5, and 1-year exposures to 10^{14} n/cm^2 sec thermal neutron flux. It was concluded that errors in W-0.5at.Os vs W-25Re could reach a maximum of +50°C after six months at 10^{14} n/cm^2 sec. This is a considerable improvement over the performance of a W vs W-25Re couple under like conditions.

W-1at.Os vs W-(25-26)Re.   Kuhlman (1966) gave a voltage-temperature curve for this couple, shown in Figure 3-16, and used the same simulation technique discussed previously to evaluate transmutation in couples of this type. The results indicated that after six months in 10^{14} n/cm^2 sec thermal flux, the error would be a little over 60°C in the 1200-1500°C range, becoming smaller at higher temperatures. Such an error would be less than that of a W vs W-25Re couple exposed to the same flux.

W-3at.Os vs W-26Re.   A calibration curve for this couple was also given by Kuhlman and is shown in Figure 3-16.

MULTICOMPONENT SYSTEMS

General Information

Here, with one exception, thermocouples are discussed for which one or both conductors are alloys of three or more elements. In some instances the third element in the ternary alloy conductor was carefully chosen and introduced for a specific purpose, such as to inhibit recrystallization. On the other hand, some of the metals such as Manganin in Manganin vs gold-cobalt were choices of convenience from existing alloys. An unusual conductor consisting of tungsten, platinum clad to improve oxidation resistance, is also described.

Ternary Alloy and Other Multicomponent Thermocouples

Pt-1W-1Mo vs Pt, Pt-2W-1Mo vs Pt, Pt-3.5Mo-1W vs Pt.   Nishimura (1958) included these couples in a patent disclosure, claiming the use of molybdenum and/or tungsten for a couple comparable with Pt-10Rh vs Pt in terms of drift, homogeneity, corrosion resistance, and heat resistance. At the same time a higher thermoelectric power was obtained. Voltage-temperature characteristics for the three thermocouples are included in Figure 3-17. The thermoelectric power is 19, 25, and 37 μV/°C for Pt-1W-1Mo, Pt-2W-1Mo, and Pt-3.5Mo-1W, respectively, at temperatures in the neighborhood of 1200°C.

Pt-5Rh-4.5Re vs Pt.   This was described by Goedecke (1931) and said by Schulze (1939) to be useful up to 1600°C. It was intended for long periods of service when in gas-tight tubes. A calibration curve from Bennett (1961, p. 27), which was attributed to Goedecke, appears in Figure 3-17. The thermoelectric power is about 27 μV/°C near 1000°C.

According to Schulze, this pair is an improvement over the Pt-8Re vs Pt couple, with the rhodium added to counteract recrystallization. The couple drifts during high temperature exposure in air, however, which Bennett demonstrated from results of his experience. He reported a loss in output of 0.7 mV at 1000°C after 3.5 hours of exposure at 1550°C.

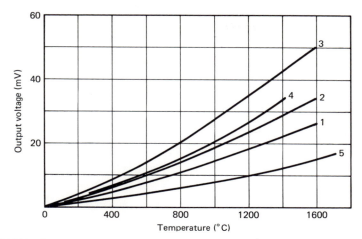

FIGURE 3-17   Multicomponent alloy thermocouples.   1. Pt-1W-1Mo vs Pt;   2. Pt-2W-1Mo vs Pt;   3. Pt-3.5Mo-1W vs Pt: Nishimura (1958).   4. Pt-5Rh-4.5Re vs Pt, Bennett (1961, p. 27).   5. Pt-2W-1Mo vs Pt-13Rh, Nishimura (1958).

Pt-2W-1Mo vs Pt-13Rh.   This thermocouple, a voltage-temperature curve for which is shown in Figure 3-17, was covered by Nishimura (1958) in his patent; the discussion under Pt-1W-1Mo vs Pt is also applicable here. The thermoelectric power is near 12 μV/°C at 1300°C, decreasing at lower temperatures.

Pt-20Rh-10W vs Pt-20Rh.   Rudnitskii and Tyurin (1956) evaluated this combination after making drift tests with each alloy versus platinum. Bare wire couples were first annealed for an unspecified time at 1600°C and then calibrated. Calibration points were taken at the freezing points of naphthalene and aluminum, and by the wire melting point method with gold, palladium, and platinum. The resulting voltage-temperature characteristic is shown in Figure 3-18. The thermoelectric power is 9.2 μV/°C at 1400°C.

Drift tests in air were conducted by the method of exposure at high temperature followed by recalibration. Tabulated information was shown giving thermocouple output after 0, 1, and 2-hour exposures at 1750-1800°C. Data for calibration temperatures of 1063, 1555, and 1773.5°C were included. Drift during the first exposure was considerably greater than the following two-hour exposure. A negative drift of -2.4 per cent can be computed from the investigators' table for the two-hour period and at the 1063°C calibration temperature. At higher calibration temperatures the drift was less in magnitude, dropping to -0.75 per cent at 1773.5°C. For comparison, there was considerably greater drift for a Pt-30Rh vs Pt-6Rh thermocouple under similar conditions; a -2.4

per cent drift for two hours at 1750-1800°C was found at the 1773.5°C calibration temperature.

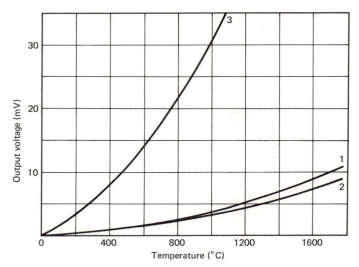

FIGURE 3-18   Multicomponent and special composite thermocouples.   1. Pt-20Rh-10W vs Pt-20Rh, Rudnitskii and Tyurin (1956).   2. Pt-20Rh-10W vs Pt-30Rh, Rudnitskii and Tyurin (1956). 3. W-42Pt(clad) vs Pd, Ihnat (1959).

It is evident that under the conditions of the test, bare wire Pt-20Rh-10W vs Pt-20Rh performed better than Pt-30Rh vs Pt-6Rh at high temperature. It is also evident that the tungsten-bearing alloy couple required a period of annealing at high temperature before it reached its best stability condition.

Couples enclosed in aluminum oxide tubes of unspecified purity showed high rates of drift at 1800°C. The drift reached 21 per cent after an exposure of 24 hours. The condition was believed by the investigators to be caused by instability of the Pt-Rh-W alloy in the insulation. This work was done before the contamination effects of iron impurities were known, so the possibility of improved performance with low iron content insulators exists.

Pt-20Rh-10W vs Pt-30Rh.   This couple was evaluated by Rudnitskii and Tyurin (1956) as described for the very similar couple in the preceding paragraphs. The voltage-temperature curve is included in Figure 3-18. The thermoelectric power is 7.1 μV/°C at 1400°C. A drift of –2.7 per cent was reported at 1773.5°C, after two hours at 1800°C. The corresponding test for Pt-20Rh-10W vs Pt-20Rh resulted in a –0.75 per cent drift.

Cu-13Mn- 4Ni-0.3Fe vs Au-1.8at.Co.   Slack (1961) chose this combination for a cryogenic application because of the low thermal conductivities of both alloys. The thermoelectric power was said to be 10.3 μV/K at 10 K, and 43.4 μV/K at 273.2 K. Slack suggested this pair for use over the full 3-300 K temperature range.[8] He referred to the copper alloy as Manganin, giving the composition shown. A nominal composition range often associated with this alloy is given by (82-86)Cu-(4-15)Mn-(2-12)Ni-Fe; the name is also associated with alloys of lower copper content and with other added elements.

Te-32Ag-27Cu-7Se-1S vs (Ag$_2$S)-50(Ag$_2$Se).   A semiconductor thermocouple with this nominal composition was described for application as an infrared detector by Brown, Chasmar, and Fellgett (1953), and Smith, Jones, and Chasmar (1957). The work was attributed to E. Schwarz and was said to be the first application of a semiconductor thermocouple as an infrared detector. The thermoelectric powers of some experimental materials prepared by Brown, et al., resulted in sensitivities that ranged from 150 to 500 μV/°C.

Au vs (Ag$_2$S)-50(Ag$_2$Se).   The thermoelectric power of this combination was reported by Brown, et al. (1953) to be 400-490 μV/°C. The construction for an infrared detector was similar to that of Schwarz, and considerable detail on radiometer theory, design, and fabrication was included.

Composite Type Thermocouple

W-42Pt(clad)[9] vs Pd.   This unusual thermocouple was described by Ihnat (1959), and Ihnat and Hagel (1960). The voltage-temperature curve is given in Figure 3-18. The thermoelectric power averages about 45 μV/°C in the 800-1000°C region. The clad wire was protected by insulation and the palladium outer conductor, present as a concentric tubing. However, this combination did not perform as well as such couples as Pt-15Ir vs Pd, during tests in the oxidizing atmosphere of a simulated combustion chamber.

THERMOELECTRIC DATA ON MISCELLANEOUS NOBLE-BASE METALS

Figures 3-19 and 3-20 give voltage-temperature curves for a number of alloys relative to platinum or other metals. The information was found in applications-oriented papers.

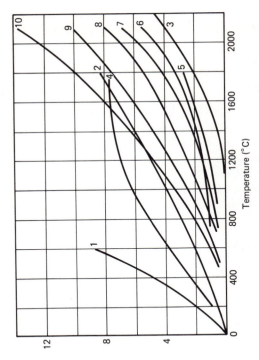

FIGURE 3-20   Additional thermoelectric voltage data on noble-base alloys, II. 1. Ag-10Cu vs Pt, Roeser and Wensel (1941, p. 1293). 2. W-25Re vs Os; 3. W-26Re vs Re-10Os; 4. Ru-10Ni vs W-25Re; 5. W-25Re vs Ru: Kuhlman (1966). 6. Ir-10Re vs Ir; 7. Ir-20Re vs Ir; 8. Ir-30Re vs Ir; 9. Ir-40Re vs Ir; 10. Ir-50Re vs Ir: Haase and Schneider (1956).

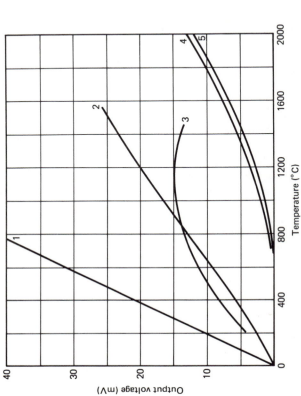

FIGURE 3-19   Additional thermoelectric voltage data on noble-base alloys, I. 1. Pt vs Au-9Ni, Hicks, et al. (1963). 2. Mo-lat.Ru vs W-25Re, Kuhlman (1966). 3. Ni-10at.Ru vs W-25Re, Kuhlman (1966). 4. Re-20Ir vs Ir, Haase and Schneider (1956). 5. Re-10Ir vs Ir, Haase and Schneider (1956).

FOOTNOTES

1. All references are located at the end of the book.
2. The voltage-temperature curves and drift characteristics of these couples may not be fully established in view of results described under Pt-5Mo vs Pt.
3. Lead telluride of unspecified doping was used, and the thermocouple polarity was not indicated.
4. The name Chromel is now rather loosely applied in the United States to a number of approximately 90 per cent nickel alloys with chromium and often with smaller amounts of other elements added. The practice developed because of wide acceptance of an alloy trademarked by the Hoskins Manufacturing Co. as Chromel-P, followed by the other thermoelectrically similar alloys that are now in use.
5. Slack used a composition of Au-1.8at.Co rather than the higher cobalt content alloy generally used.
6. This behavior can be explained in the same general way as the behavior of Ir-40Rh vs Ir, which was discussed earlier. The iridium oxidizes more rapidly than rhodium in that case. Here the rhenium oxidizes faster than the iridium.
7. The specimen for which data were given showed a two-phase microstructure. According to Kuhlman the output was slightly negative with respect to W-26Re.
8. Thermocouples using gold-cobalt alloys are metallurgically unstable. This is discussed under Ag-0.37at.Au vs Au-2.1at.Co.
9. This is not an alloy, but a tungsten wire clad with platinum, the cladding being 42 per cent by weight of the total composite material.

4

Base Metal
Thermocouples for
Low and Moderate
Temperatures

COMMONLY USED MATERIALS

Certain base metal elements and alloys are often used in thermocouple applications. General information on some of these metals is given below for use in later sections.

Copper

Good quality copper has unusually desirable properties for use either in thermocouples or as a standard material for thermoelectric comparison purposes. Commercial copper that meets specifications for soft or annealed bare copper wire can be used for thermocouples without special selection processes. Annealed electrolytic copper makes an excellent standard for thermocouple testing of materials below 300°C and is quite comparable with platinum for this purpose.

Commercial copper is very homogeneous and of high uniformity. Further, mishandling that causes severe cold working does not produce the large thermoelectric changes which result in most other metals when similarly treated. The thermal conductivity of the metal is high, which may or may not be an advantage depending on the application. The only important drawback of copper is its tendency to oxidize rapidly at high temperatures. This limits its use in air to temperatures below about 400°C. Because of the oxidation problem small diameter bare copper wire has a much shorter useful life than that of larger sizes.

117

Iron

Iron is not economically available in as pure a state as copper. Therefore, thermo-couple iron is selected by choosing iron which closely matches the thermoelectric properties of the iron used in establishing the earliest iron-constantan thermo-couple table in 1913. The exact composition is not specified, so some variations are encountered from one supplier to another. Thermocouples using a pure iron such as electrolytic iron cannot be expected to match the output of commercial thermocouples.

Cu-(35-50)Ni

The name constantan is associated with alloys of copper and 35 to 50 per cent nickel. These alloys have been manufactured under such trade names as Advance, Copel, Copnic, Cupron, Eureka, Excelsior, Ferry, Ideal, and Thermoconstantan. For thermoelectric purposes the composition of the alloy generally employed in the United States is approximately Cu-43Ni; sometimes small amounts of man-ganese and iron may be present, as well as trace impurities. Other compositions have also been used, particularly in Europe.

Finch (1962)[1] outlined the development of constantan for thermocouples. This alloy was originally used where a low temperature coefficient of electrical resistance was needed. Later it was adopted for thermoelectric use to take ad-vantage of the high thermoelectric power and other favorable properties of the copper-nickel alloys. Although the nominal composition of Cu-43Ni was the most suitable for its original application, it was not the composition of greatest thermoelectric power. Nevertheless, it was put into use for thermocouples be-cause it was available, and is now firmly established for application with the commonly used base metal couples. Therefore, in special applications the opti-mum alloy composition of about Cu-40Ni could be used and some improvement would be obtained.

Constantan is selected for use with copper, or with iron previously selected for thermocouples, on the basis of a good thermoelectric match of thermocouple output with the appropriate table. There is no specification for the exact compo-sition. Also, because of differences in the tables now in use, constantan selected for use with copper is unacceptable for use with iron, with regard to obtaining a closely matched relation with the reference tables.

Ni-10Cr, Ni-3Mn-2Al-1Si

The alloys Chromel-P and Alumel have the respective nominal compositions given above. The use of these and similar alloys began with the Chromel-Alumel thermocouple and is discussed in detail under that heading.

Ni-(14-20)Cr

Nickel alloys with about 18 to 20 per cent chromium, and usually with two or more other elements in smaller amounts came into use during the mid 1950's because of superior oxidation resistance and improved characteristics in some reducing atmospheres. Lower nickel content alloys such as Ni-14Cr were also studied. Although use of these alloys in the United States has been relatively recent, the desirable properties of a similar alloy, Ni-16Cr, were described earlier by Rohn (1927). There has also been interest in a nickel alloy with about 15 per cent chromium plus a small amount of silicon, because of the possibility of improved oxidation and stability characteristics. This was suggested by Burley (1971).

THERMOCOUPLES CONTAINING COPPER

Elemental Combinations

Cu vs Fe. The voltage-temperature characteristic is given in Figure 4-1, plotted from data by Roeser and Wensel (1941, p. 1293). It can be seen that at high temperatures copper is positive, while at and below an intermediate temperature, which depends on the iron impurity content, iron becomes the positive metal.

Kuether (1960) took advantage of the special properties at the intermediate temperature for use in difference temperature measurements in a copper block. In a different type of application, Danishevskii and others (1964) reported that iron and copper can be employed as extension wires with the W-10Re vs W-20Re couple, which is often used in Russia for high temperature work.

Marshall and coworkers (1966) developed thin film couples consisting of vapor deposited metals on glass slides. The slides were heated during deposition to obtain films in the annealed state, and after deposition the films were coated with a layer of silicon monoxide to provide protection from corrosion and abrasion. Each couple was calibrated in the 20-300°C range.

Cu vs Ni. The voltage-temperature curve, from Roeser and Wensel (1941, p. 1293), is plotted in Figure 4-1. The thermoelectric power is 27 μV/°C near 200°C. Base metal pairs using alloys of copper and nickel are superior to this combination in terms of such properties as output and oxidation resistance, but Cu vs Ni has been popular for use in vapor deposited or electroplated couples. Vapor deposition is relatively simple with pure elements; there are no inadvertent changes in composition, which sometimes occur with alloy deposition.

Watson (1966) made a review of surface temperature measurement methods and his earliest reference for this couple was that of Othmer and Coats (1928), who applied a nickel film over a short section of copper tube. Also mentioned was later work by Hyman, Bonilla, and Ehrlich (1951) who electroplated overlapping films of copper and nickel on a quartz cylinder to form the junction.

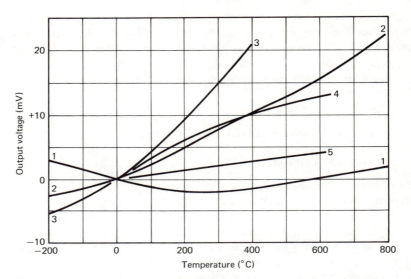

FIGURE 4-1   Thermocouples with copper.   1. Cu vs Fe, Roeser and Wensel (1941, p. 1293).   2. Cu vs Ni, Roeser and Wensel (1941, p. 1293).   3. Cu vs Cu-43Ni, Shenker, et al. (1955).   4. Cu vs Fe-28Ni-18Co, Rauch (1954).   5. Ni-(14-17)Cr-(6-10)Fe vs Cu, Rauch (1954).

Other thin film work was done by Reid and Gross (1957), who utilized a very thin strip of bamboo as the substrate in an application where a rapid response time was desired. Thermocouples with one or both conductors of thin films were described by Behrndt (1960), Bullis (1963), Marshall, et al. (1966), and Moeller, et al. (1968).

Rauch (1954) used an insulated copper wire inserted into small diameter nickel tubing for a needle thermocouple application. The thermoelectric power was 25 $\mu V/°C$ in the 500°C region. In spite of the interest in this thermocouple there is no information available on experiences with drift or atmospheric effects.

Te vs Cu.   A difference thermocouple was described by Hatfield and Wilkins (1950), consisting of a Cu-Te-Cu sandwich of metals, for a heat flux sensor application. A useful temperature range of up to 200°C was mentioned. Considerable drift was reported after lengthy heating.

Alloy versus Element Pairs

Al vs Cu-43Ni.[2]   This combination was used by Cook and McCampbell (1968) to control the temperature for welding aluminum alloy plates. The plate formed one thermocouple conductor while a constantan probe, held against the plate under pressure, acted as the other conductor. The thermoelectric power was said to be about 38.5 $\mu V/°C$. Repeatability of a plate surface temperature measurement appears to have been within 2 per cent.

Cu vs Cu-(0.1-0.3)Fe.   Medvedeva and coworkers (1971) in their conference paper described the properties of copper-iron alloys in the 2 to 300 K temperature region. They suggested this combination as a cryogenic thermocouple. The thermoelectric power was at least 14 $\mu V/K$ in the 4 K region and was about 5 $\mu V/K$ at room temperature.

Cu vs Cu-1.78Ni.   Extension wires of this type were mentioned by Danishevskii and associates (1964) for use with the W-5Re vs W-20Re thermocouple.

Cu vs Cu-3Ni.   According to Fanciullo (1964) this pair is commercially available for use as extension wires with Pt-10Rh vs Pt couples. It was also selected by Fanciullo as the best possibility for extension wire use with Mo vs W-26Re and Mo vs Nb.

Cu vs Cu-43Ni.   Copper-constantan thermocouples have been used for many years. A calibration table published by Adams (1914) became the first widely used reference table, commonly referred to as the Adams Table. In the 1930's, calibration work at the National Bureau of Standards resulted in the reference tables by Roeser and Dahl (1938). These results have sometimes been referred to in the literature as the 1938 table of Adams; the grade of constantan which met the thermoelectric output of the table, when paired with good quality copper, was called Adams constantan.

Later, Shenker, Lauritzen, Corruccini, and Lonberger (1955) developed revised reference tables, based on those of Roeser and Dahl for temperatures above 0°C and upon work by Scott (1941) for temperatures below 0°C. These are included in the well-known National Bureau of Standards Circular No. 561. A voltage-temperature curve from this work is shown in Figure 4-1. The thermoelectric power ranges from 38 $\mu V/°C$ at 0°C to 61 $\mu V/°C$ at 370°C. Jones (1969) converted the NBS tables to the International Practical Temperature Scale of 1968 for the range of –180 to +390°C in one degree increments. Gottlieb (1971) discussed converting to the new scale and also gave conversion data.

Manufacturers of thermocouple wire have followed the practice of tailoring the choice of the constantan to maintain a copper-constantan output close to the tabulated values, when used with a good grade of copper. Temperature tolerance limits are then given for the paired wires. Therefore there are some variations in the actual alloy composition. Because of the original choice of alloys from available materials, and because of the trends of thermocouple usage, constantan which is at present acceptable for copper-constantan use is not acceptable with Type J iron-constantan, and vice versa.

The range and accuracy tolerances typical of commercial thermocouple wires are given in Table 4-1, primarily from Starr and Wang (1969). Both standard and special or premium grades of wire are available for most ranges. Manufacturers and interested organizations have adopted type numbers for the most used couples, and copper-constantan couples that closely match the NBS Circular 561 tables are labeled Type T.

TABLE 4-1

Recommended Ranges and Representative Accuracies of Commercial Copper-Constantan Thermocouples

Temperature Range (°C)	Tolerance Limits in Per Cent of Reading or in Celsius Degrees	
	Standard	Special
−190 to − 60		±1.0%
−100 to − 60	±2.0%	±1.0%
− 60 to +100	±0.8°C	±0.4°C
+100 to +370	±3/4%	±3/8%

Most discussions of this thermocouple give a maximum temperature rating which varies from 300 to 370°C. Billing (1963) estimated an accuracy figure of 10 per cent at 500°C. The limiting factor here is the rapid oxidation of the copper at high temperatures. Fine wire thermocouples have a short lifetime in air at high temperatures whereas heavy gage wires last longer. Whether the wires are bare, enclosed, or in packed insulation is also important. The maximum temperature in inert atmosphere or vacuum is considerably higher than the limit in air. Unfortunately, there is little quantitative information on high temperature limitations in spite of the large number of uses to which this couple has been applied.

Billing also stated that, once individual calibrations are made, a repeatability of ±0.1°C is possible over the −200 to +200°C interval. Further, an equation relating voltage to temperature can be fitted to the output curve within that same ±0.1°C, an unusual degree of matching for base metal couples.

Roeser and Lonberger (1958) indicated that a given couple will maintain its calibration within ±0.2°C up to 300°C. The voltage-temperature characteristic can be represented by an equation of form

$$V = aT + bT^2 + cT^3$$

between 0 and 300°C, to within ±0.2°C. The constants a, b, c, are determined by a set of three calibration points at about 100, 200, and 300°C. A similar relation holds for 0 to –190°C. Between 0 and 100°C a simpler equation of form

$$V = aT + 0.04T^2$$

can be used within ±0.2°C limits. In this equation V is in microvolts and T is in degrees Celsius; the constant a is determined by a single calibration at 100°C. Bondareva (1969) described an interesting method of calibrating in the 0 to –200°C range by means of one calibration point at +100°C.

When these thermocouples are compared with the reference tables, or with a calibrated reference thermocouple, the calibration output differences are usually proportional to the thermoelectric voltage. In other words, a 10 μV difference at 1 mV output will double when the output is 2 mV. Scott (1941) showed difference curves for 17 couples and 2 calibration tables between 0 and –200°C and most of these curves were approximately linear. A few showed unusual increases below about –150°C. Efremova and coworkers (1963) investigated this relationship for a number of different constantan specimens over a wide temperature range and encountered only one exception to the rule.

The latter workers also investigated deformation effects on voltage and found that constantan was considerably more sensitive than copper. Dauphinee, MacDonald, and Pearson (1953), and Pollack and Finch (1962) reported similar results. Work by Powell, Caywood, and Bunch (1962) indicated that between 4 and 300 K copper-constantan is generally more homogeneous than iron-constantan, Chromel-Alumel, or Ag-0.37at.Au vs Au-2.1at.Co. Tests by Bragin and Tetyueva (1964) with Russian made alloys showed that copper-constantan had less inhomogeneities in the –200°C region than Chromel-Alumel or Chromel-constantan.

For general applications, copper-constantan can be used in reducing, oxidizing, or inert atmospheres as well as in vacuum. The higher temperature range is rather restricted, however, compared with iron-constantan. On the other hand, copper-constantan can be used in moist atmospheres and at temperatures below 0°C without significant corrosion; this is an advantage over iron-constantan.

The copper-constantan thermocouple was for many years one of the most important combinations for cryogenic temperature measurements in the 11 to 300 K range. It has been supplanted to some degree for temperatures below about 80 K by the gold-iron alloys, used in conjunction with normal silver or other metals. For many applications it remains a good choice above 80 K, particularly where temperatures must also be measured up to 370°C. One reason for the decreased usage of copper-constantan below 80 K is the low thermoelectric power S, compared with more recently developed cryogenic couples. For example, at 10 K, S is about 3 μV/K compared with 10 to 15 μV/K for the newer types. Further, the newer couples can also be used to 4 K and lower, a range in which copper-constantan can hardly compete. Finally, the high thermal conductivity of the copper is sometimes a serious disadvantage at these temperatures because of the possibility of heat leakage.

Calibration work for the cryogenic range down to as low as 11 K was done by W.F. Giauque and others in the 1920's. Similar efforts by Aston, Willihnganz, and Messerly (1935) resulted in a calibration table for the 12-90 K interval. Other cryogenic work with this thermocouple was reviewed by Scott (1941), and applications data were given by Aston (1941). Studies of homogeneity and applications were made by Fuschillo (1954). More recently, several cryogenic calibration and application studies have been described: Powell, Bunch, and Caywood (1961), Powell, Caywood, and Bunch (1962), Corruccini (1963), Sparks, Powell, and Hall (1968), Bondareva (1969), and Richards, et al. (1969).

In other work, Marshall and associates (1966) studied thin film couples deposited on glass substrates, with a silicon monoxide coating to provide film protection. The alloy was flash evaporated onto the glass. The thickness of the copper was 5000 Å or less; that of the constantan was 1000 to 2000 Å. Output of the couple was lower than that of the corresponding bulk metals. The investigators suggested that a thicker alloy film might have improved the output.

These couples could be used up to 200°C or more. Curves comparing a freshly prepared couple with one that had been baked out for 15 minutes at 450°C demonstrated very similar outputs in the 240°C range. Nevertheless, at lower temperatures there were differences, showing that the bakeout had affected the voltage-temperature characteristics.

Work by Chopra and fellow investigators (1968) resulted in couples with higher outputs. These indicated nearly bulklike behavior when the thickness of both elements was greater than some 1200 Å. A response time of less than a microsecond to a pulsed beam laser source was reported.

Baxter, et al. (1969) found that small wire diameter thermocouples with magnesium oxide insulation and stainless steel sheaths performed exceptionally well during thermal cycling tests between 316 and 649°C. Several thermocouples were tested for 1387 and 1625 thermal cycles without failures; this performance was attributed to a relatively good match between the thermal expansion coefficients of the materials in the cables. Other types of thermocouples exhibited failures in some cases.

Palmer and Turner (1966) reviewed earlier work by themselves and by others on shock sensitivity and described shock test results for pressures of 50 to 400 kB. Transient output voltages were observed which were much higher than the expected outputs due to shock-induced temperature rises.

Dau and coworkers (1968) reviewed the literature covering nuclear heating of materials. They referred to early work by Boorman (1950), Madsen (1951), and Jamison and Blewitt (1958), with copper-constantan couples, and to other work with different metals. From this information it was concluded that reported nuclear dose rate effects on thermocouple calibration result in less than 1 per cent output change for flux of less than 10^{13} n/cm^2 sec. Transmutation effects were not discussed.

Bundy (1961) described early work by Bridgeman (1918) concerning the effect of high pressure on thermocouple output, and then described his own experiments at high pressures. Bundy found that increased pressure resulted in a decreased output for copper-constantan. A nearly linear temperature vs pressure correction curve was shown from 0 to 70 kB, giving an additive correction of about 0.052°C/kB, when there was a 100°C temperature difference between the thermocouple junctions. The junctions for Bundy's experiments were maintained near 0 and 100°C. Bloch and Chaissé (1967) reported on work which supported Bundy's results; it appears that between –196 and +89°C the voltage decreases by (2.2 ±0.25) μV/kB for a temperature difference of 100°C between the thermocouple junctions.

Cu vs Cu-0.005Sn.   Dauphinee, MacDonald, and Pearson (1953) suggested the use of a very dilute alloy of tin in copper, based on study of Cu-0.0066Sn and other alloys in the 0 to 70 K range. They were of the opinion that a Cu vs Cu-0.005Sn couple could be used from 2 to 30 K. Pearson (1954) further discussed thermocouples using dilute alloys of copper, which have good sensitivities and homogeneity for the cryogenic region. At the same time, these alloys have practically zero thermoelectric power at the ice point provided that the solute concentration is not too high. Although these advantages are desirable properties for cryogenic applications, the high thermal conductivity of these metals is often undesirable for such temperatures because of heat leakage. Also, Pearson indicated that dilute alloys of specified composition are difficult to prepare because of the sensitivity to contamination.

Cu vs Cu-(3.5-8)Sn-(0.05-0.5)P.   The combination of copper versus phosphorus-bronze was used by Holtby (1941) for extension wires with the W vs C (graphite) thermocouple. No alloy composition was given. The composition given here is a nominal composition range for phosphor-bronze spring wire; this alloy has been discussed by Woldman (1949, p. 158). These metals were used as extension wires because their output as a thermocouple was the same as that of tungsten versus spectroscopic grade graphite, up to 150°C. The thermocouple voltage at that temperature was on the order of 0.5 mV.

Cu vs Fe-(18-20)Cr-(8-12)Ni-(0-2)Mn.[3]   Zysk and Robertson (1971) mentioned the use of copper vs Type 304 stainless steel for extension wires with thermocouples of iridium 40, 50, or 60 per cent rhodium versus iridium. The extension wires were useful up to 200°C but calibration of each lot of wire was necessary.

Cu vs Fe-(16-18)Cr-(10-14)Ni-(2-3)Mo-(0-2)Mn.[3]   The alloy is Type 316 stainless steel. Rauch (1954) described this pair, and others, for a needle thermocouple application. The stainless steel conductor was a 2.1 mm diameter tube containing an insulated copper wire. The hot junction was established by fusing the wire and the tube together at the end. The unit was calibrated by comparison with a Chromel-Alumel couple.

Cu vs Fe-(17-19)Cr-(9-13)Ni-(0-2)Mn.[3]   Here the alloy is Type 347 stainless steel. Moeller and co-authors (1968) described the use of this pair as extension wires for the Ir-40Rh vs Ir thermocouple. The voltage-temperature characteristics of the two combinations did not match precisely. There were also spool-to-spool variations, probably due to the Type 347 alloy.

Cu vs Fe-28Ni-18Co.   Rauch (1954) included copper vs Kovar[4] with other pairs for a needle thermocouple application. An insulated copper wire was inserted in a 1 mm diameter Kovar tube. The hot junction was established by fusing the wire and the tube together at the end. Calibration was by comparison with a Chromel-Alumel thermocouple. The curve is quite nonlinear and is shown in Figure 4-1. The thermoelectric power is about 13 μV/°C near 500°C.

Ni-(14-17)Cr-(6-10)Fe vs Cu.   Rauch (1954) prepared and calibrated an Inconel-copper couple in the same manner as was done with copper vs Kovar, with the exception that 1.8 mm Inconel tubing was used. The alloy was referred to only by name; the nominal composition generally given is used here, and there may also be small amounts of manganese and other elements present. The output curve is given in Figure 4-1. The thermoelectric power is low, about 7 μV/°C at 500°C. The couple showed erratic output during recalibration; no details were given on this behavior.

Fe vs (82-86)Cu-(4-15)Mn-(2-12)Ni-Fe.   Studennikov and Erkovich (1965) reported on the use of this combination for extension wires with W-5Re vs W-20Re. The alloy was specified as Manganin which met the Russian specification PÉShOMM. No composition was stated so a nominal composition range[5] is given here.

Both unannealed and annealed iron wires were evaluated. Annealed iron paired with Manganin proved to most closely match the tungsten-rhenium alloy thermocouple characteristics, and was recommended for extension wires in the 10 to 50°C temperature range. The output difference between the extension wire pair and W-5Re vs W-20Re was reported to be 3 per cent or less in that interval. At higher temperatures the output of the extension wires was low, 13 per cent below that of the tungsten-rhenium couple at 90°C.

Fe vs Cu-43Ni.   The early history of this thermocouple was outlined by Finch (1962). During the initial use of this couple in the 1910 era, there was a vigorous opposition to the application of iron for thermocouples, because the iron available at that time was quite inhomogeneous. Nevertheless, the high thermoelectric power of this couple, a comparatively low cost, and an adaptability to both oxidizing and reducing atmospheres made it an attractive choice for industrial applications. Since that time there has been much improvement in the quality of the iron.

The earliest calibration table dates back to 1913, and this was only slightly modified by Corruccini and Shenker (1953) when they developed reference tables for the couple. This work was later incorporated by Shenker, et al. (1955) in NBS Circular No. 561. The latter reference tables are currently used by manufacturers to match selected iron and constantan pairs within specified tolerance limits and obtain the Type J thermocouple curve. Type letters are in general use by industry and interested organizations in the United States to designate the more commonly used thermocouples. From data by Starr and Wang (1969), thermocouples are commercially available with a standard tolerance of ±2.2°C from 0 to 280°C and ±3/4 per cent from 280 to 760°C. Special or premium grade wires are also available with tolerances of ±1.1°C and ±3/8 per cent for the respective ranges of 0-280 and 280-760°C.

A voltage-temperature curve from the data in NBS Circular 561 is shown in Figure 4-2. At -100°C the thermoelectric power S is about 40 μV/°C. At 0, +100, and +500°C, S is 50, 54, and 56 μV/°C, respectively. Jones (1969) converted the NBS tables to the International Practical Temperature Scale of 1968 for the range of -180 to +750°C in one degree increments. Gottlieb (1971) discussed converting to the new scale and gave conversion data.

During the earlier years of thermocouple usage, somewhat different calibration tables were developed for some applications, particularly for military aircraft. The current reference tables for this version of the iron-constantan pair were given by Roeser and Dahl (1938) and the couple is now designated as Type Y. The constantan which was used was Adams constantan, used also for the copper-constantan thermocouple; selected samples of iron were chosen to represent the positive conductor. There are some small differences in the voltage-temperature characteristics of the Types Y and J, but not in their other physical properties.

The normal temperature range for iron-constantan is -190 to +760°C, with a maximum temperature usually given in the 900-1000°C range. It has also been used to some degree well below -190°C. Powell, Caywood, and Bunch (1962) gave a calibration table for the cryogenic range. At 10 K, S is 3.3 μV/K, comparable with copper-constantan for which S is 3.1 μV/K. Inhomogeneities in Fe vs Cu-43Ni are not as small as for copper-constantan, however, and Powell, et al., recommended Cu vs Cu-43Ni as a better choice.

Calibrations can be made by comparison with a standard thermocouple at approximately 100°C intervals between 0 and 760°C, and the resulting uncertainty of interpolation temperatures within that interval is 1°C, according to Roeser and Lonberger (1958). The uncertainty can be improved to about 0.5°C over the 0-350°C interval, but this still does not compare too well with copper-constantan between 0 and 300°C for which the corresponding uncertainty is 0.2°C.

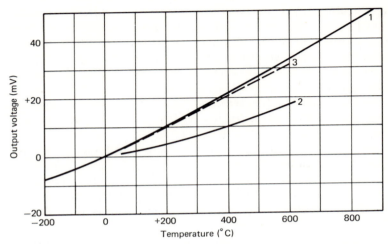

FIGURE 4-2   Thermocouples with constantan, I.   1. Fe vs Cu-43Ni, Shenker, et al. (1955).   2. Ni vs Cu-43Ni, Rauch (1954). 3. Zr vs Cu-43Ni, Shoens and Shortall (1953).

Little information has been published regarding the use of equations to approximate the voltage-temperature tables for temperatures above 0°C, although an equation for the 0 to –190°C range was given by Finch (1962) with the form

$$V = aT + bT^2 + cT^3$$

White (1968) analyzed the tabulated data for several thermoelectric materials versus platinum given by Roeser and Wensel (1941, p. 1293), and produced fourth degree equations for iron and constantan, relative to platinum. These equations were intended for the 100-1000°C interval. It was stated that when used to compute the output of Fe vs Cu-43Ni, the error relative to Roeser and Wensel's work ranged from 0.08 to 0.56 per cent, the latter figure corresponding to the 1000°C point.

Iron-constantan can be used in oxidizing, reducing, or inert atmospheres as well as in vacuum. Heavy gage wires can be briefly used up to 1000°C, although the normal limit is 760°C as mentioned earlier. There are few thermocouples that can tolerate reducing atmospheres, so iron-constantan is particularly important in this respect. Sulfurous atmospheres limit the upper temperature to about 550°C. Iron rusts in moist atmospheres and both rust and embrittlement may occur below 0°C.

Studies of long term effects in air were made by Dahl (1941). These studies were probably the first comprehensive evaluations of thermocouple drift caused by environment. It was found that both iron and constantan, relative to a common platinum reference electrode, showed drift during high temperature exposure to air. The change in the case of the constantan was gradual and cumulative. For iron the change was small until failure of the wire was approached. The change then became rapid and relatively large. Nevertheless, iron and constantan wires of the same size were found to have very similar useful lifetimes.

A list of drift test results for Fe vs Cu-43Ni is given in Table 4-2. Wire diameter affected the drift. At 760°C a 3.26 mm diameter wire pair changed by about -2°C after 400 hours. A 1.02 mm diameter pair changed by about -10°C under the same conditions. For a given drift test temperature, the change in calibration was often higher at higher calibration temperatures, but no well behaved relationship existed in this respect.

TABLE 4-2

Drift of Iron-Constantan Thermocouples in an Atmosphere of Air

Temperature (°C)	Exposure Time (Hours)	Wire Diameter (mm)	Maximum Apparent Temperature Change (°C)
427	1000	1.63	< 0.6
538	1000	1.63	< 0.6
649	1000	3.26	− 2.2
760	800	3.26	− 3.9
871	100	3.26	− 5.6
982	28	3.26	−10
1093	8	3.26	−10.6

Immersion tests were also conducted by Dahl, and some very noticeable changes of thermocouple output occurred when the depth of immersion was changed after 20 hours at constant temperature. A couple heated in air for 20 hours at 871°C showed a -4.9°C change in output at that temperature when withdrawn 7.6 cm from its original location.

Evans and Holt (1967) described drift of a sheathed couple during exposure in an atmosphere of carbon dioxide. The couple was insulated with magnesium oxide and in a stainless steel sheath, with an insulated hot junction. After 190 days at 400°C a positive drift of 0.25°C was observed, which was considered as negligible.

Cold work experiments were conducted by Pollock and Finch (1962), who found that almost all output changes occurred immediately after cold working.

Both iron and constantan wires were shown to be somewhat susceptible to cold-work-induced changes in output. Drawing of metal sheathed insulated thermocouple cables during manufacture may produce changes in output characteristics. Schwiegerling (1971) described observations where the output generally decreased during several drawing and intermediate annealing steps, and eventually the output error exceeded acceptable limits.

Thin film thermocouples of iron-constantan were prepared by Marshall and coworkers (1966), where a flash evaporation process was used to deposit the alloy and obtain the proper film composition. Glass slides were used as substrates and were heated during the deposition process to obtain films in the annealed state. An over-coating of silicon monoxide was deposited on the films to provide protection from corrosion and abrasion. Thin film couple calibrations showed lower outputs than bulk metal couples. The alloy film thickness was never more than 2000 $\overset{\circ}{A}$, and the authors suggested that a thicker film might have resulted in a higher output.

A few nuclear environmental results have been reported for this thermocouple. Jamison and Blewitt (1953) exposed couples to nuclear radiation during liquid nitrogen bath immersion and found no sensitivity to an integrated flux of 5×10^{16} n/cm^2. Bianchi and associates (1966) later reported conditions where a total integrated flux of 4×10^{20} n/cm^2 was accumulated at 650°C. No transmutation induced drift was observed within the limits of the experimental error, which were about 2 per cent. An undetermined amount of nuclear heating was observed. Leonard and Hunkar (1967) observed a temporary change of –9°C in thermocouple calibration at 100°C, which was present about 2 hours after exposure to a fast neutron flux of 6.5×10^{12} n/cm^2 sec. Leonard (1969) discussed the various effects observed with this and other thermocouples in a nuclear heating environment.

The effects of high pressure on thermocouple output were described by Hanneman and Strong (1965). A curve of the absolute correction for measured temperature was shown as well as a relative correction of iron-constantan against a Pt-10Rh vs Pt thermocouple, all at 40 kB. The relative correction curve shows a pronounced change in slope, corresponding to the α to γ transformation in iron at 685°C and 40 kB.

Ni vs Cu-43Ni.   Rauch (1954) prepared and calibrated this combination for a needle thermocouple application. The work was done in a manner similar to that for Cu vs Fe-28Ni-18Co, except that the finished nickel tube was 1.2 mm in diameter, and an insulated constantan wire was contained in the tube. The calibration curve is shown in Figure 4-2. The thermoelectric power is 38 μV/°C near 500°C.

Zr vs Cu-43Ni.   Thermocouples utilizing zirconium were studied by Shoens and Shortall (1953) for the measurement of surface temperatures of zirconium-clad fuel elements for nuclear reactors. It was desired to introduce as little extraneous material as possible into the path of the coolant fluid or the neutron flux, so zirconium was evaluated for use as one of the thermocouple materials.

A calibration was made between 100 and 450°C, using the boiling points of water and four other substances. Generally, boiling point determinations were repeated four times and barometric correction formulas were used. The zirconium was analyzed and was found to have total impurities of less than 0.5 per cent. Iron, magnesium, and silicon were present in the 0.01-0.1 per cent range; a number of other elements were present in the 0.01 per cent range.

A voltage-temperature equation was developed from the boiling point data and used to construct a calibration table. The curve in Figure 4-2 is plotted from the table. It can be seen from the figure that Zr vs Cu-43Ni and Fe vs Cu-43Ni are very similar. Shoens and Shortall pointed this out and showed that above 100°C the slopes of the two curves are very nearly the same. From the derivative of the voltage-temperature equation, the thermoelectric power for Zr vs Cu-43Ni can be shown to equal 56.2 and 52.9 μV/°C at 100 and 400°C, respectively.

Thermocouples with Two Alloys

Bi-11Sb vs Cu-43Ni. Cartwright (1933) chose this pair for use in a radiation thermopile cooled to 90 K. The bismuth-antimony alloy had a high positive thermoelectric power at that temperature.

Cu-3Ni vs Ni-3Mn-2Al-1Si. A copper-nickel vs Alumel combination was selected for extension wire use with W-5Re vs W-26Re by Fanciullo (1964). This pair matched the tungsten alloy couple better for temperatures up to 120°C than the wires commercially available at that time.

Cu-12.3Ni vs Cu-20Ni. Schulze (1939) briefly reviewed this couple, which had been developed by Goedicke (1931) to match the output of Pt-10Rh vs Pt. Good agreement was reported up to 900°C. Neither alloy is as stable as constantan.

Cu-1.25Ni-0.74Mn-0.01C vs Ni-45Cu. Zysk (1968) gave these compositions for extension wire use with (55-83)Pd-(14-34)Pt-(3-15)Au vs Au-(35-40)Pd; the compositions of Platinel I and II thermocouples fall within this category.

(82-86)Cu-(4-15)Mn-(2-12)Ni-Fe vs Cu-43Ni. Manganin vs constantan was used by Hukill (1941) in a thermocouple anemometer. The advantage of Manganin over copper was due to its much lower thermal conductivity, which permitted lower power for heating the thermocouple junction. At the same time, replacement of copper in copper-constantan by Manganin did not appreciably change the thermocouple output. The voltage-temperature curve of this combination, from data on both metals relative to platinum from Roeser and Wensel (1941, p. 1293), appears in Figure 4-3.

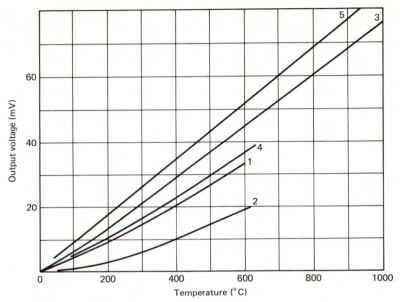

FIGURE 4-3   Thermocouples with constantan, II.   1. (82-86)Cu-
(4-15)Mn-(2-12)Ni-Fe vs Cu-43Ni, Roeser and Wensel (1941, p.
1293).   2. Fe-28Ni-18Co vs Cu-43Ni, Rauch (1954).   3. Ni-10Cr
vs Cu-43Ni, Shenker, et al. (1955).   4. Ni-(14-17)Cr-(6-10)Fe vs
Cu-43Ni, Rauch (1954).   5. Ni-16Mo vs Ni-50Cu, Schulze (1939).

Fe-5Cr vs Cu-43Ni.   Kellow and coworkers (1969) described a thermocouple
where the iron alloy was designated as a 5 per cent chromium hot working tool
steel, but the composition was not further specified. Thermocouple polarity
was not stated, but is likely to be as shown, since most high iron content alloys
are positive relative to platinum, while constantan is one of the most negative
base metal alloys. Surface thermocouples of this composition were used in measuring transient surface temperatures of dies during the forging of steel billets. A
review of work by others on this type of thermocouple installation was given by
the authors, including a steel vs constantan design used earlier by Vigor and
Hornaday (1962).

Fe-28Ni-18Co vs Cu-43Ni.   A couple of this composition was described by
Rauch (1954), and was included with a number of others previously described
for use in a needle thermocouple application. The composition shown here was
given for Kovar, the positive alloy.[6] The thermocouple consisted of an insulated

constantan wire which was inserted in a 1 mm diameter Kovar tube; the hot junction was established by fusing the wire and the tube together at the end.

Calibration was by comparison with a Chromel-Alumel thermocouple up to 635°C; two heating and cooling cycles were applied. The curve is included in Figure 4-3. The thermoelectric power is near 45 $\mu V/°C$ at 500°C.

Ni-10Cr vs Cu-43Ni.   Nominal compositions are given here for both Chromel and constantan. This thermocouple has been known for many years as the combination with the highest output of any of the commonly used metals, yet it has been little used compared with Chromel-Alumel, copper-constantan, and iron-constantan. The reference tables for Chromel-constantan were developed by Shenker and coworkers (1955) and given in NBS Circular 561. The tables are based on Chromel-platinum data, used for earlier Chromel-Alumel tables by Roeser, Dahl, and Gowens (1935), and on constantan-platinum data by Roeser and Dahl (1938), from the work leading to the tables for copper-constantan and Type Y iron-constantan thermocouples. Gottlieb (1971) gave data for converting the NBS Circular 561 tables to the International Practical Temperature Scale of 1968.

A curve of voltage versus temperature is shown in Figure 4-3 from the work by Shenker, et al. The thermoelectric power S increases from 58.5 $\mu V/°C$ at 0°C to 81.0 $\mu V/°C$ in the 400-600°C region. At 1000°C S declines somewhat to a value of 75.0 $\mu V/°C$.

Chromel-constantan wires are commercially available as Type E materials. The manufacturers supply the wires after selecting for a match within certain limits or tolerances relative to the reference tables. Representative information on tolerances was summarized by Starr and Wang (1969). Between 0 and 320°C the tolerance is ±1.7°C or ±1.25°C for standard or special grades, respectively. Between 320 and 870°C the tolerances are ±1/2 per cent and ±3/8 per cent for the two grades. In some cases brief use to 1100°C is possible. Except for lower temperature limits, usage restrictions in most atmospheres are the same as for Chromel-Alumel thermocouples.

Wintle and Salt (1967) reported that when high accuracy is desired and individual calibrations of each thermocouple are made, deviations from the smooth curve drawn through a set of calibration points are smaller for Chromel-constantan than for Chromel-Alumel type thermocouples. Wintle and Salt advised against the use of this couple in reducing, carburizing, or sulphurous atmospheres because of the resulting embrittlement and changes of thermoelectric properties.

The thermocouple performs well in oxidizing atmospheres, although it is somewhat more limited than Chromel-Alumel in temperature range. According to Dahl (1941), the limiting factor is the constantan because the latter has a higher oxidation rate than Chromel. Dahl conducted drift tests of Chromel and constantan wires in clean air, referenced to platinum. It was found that the thermoelectric changes in the two types of wire were considerably greater than the drift of completed couples of the two metals. This was because the drifts were in opposite directions so that a partial cancellation or compensation for drift resulted in the complete couple. A summary of the drift test results is given in Table 4-3.

TABLE 4-3
Drift of Chromel-Constantan Thermocouples in an Atmosphere of Air

Temperature (°C)	Exposure Time (Hours)	Wire Diameter (mm) Chromel	Constantan	Maximum Apparent Temperature Change (°C)
427	1000	1.02	1.63	<0.6
538	1000	1.02	1.63	<0.6
649	1000	3.26	3.26	−0.6
760	1000	3.26	3.26	−1.1
871	100	3.26	3.26	−2.2

Northover and Hitchcock (1968) conducted stability tests of several different Chromel and constantan alloys and also found a tendency of one to compensate for deviation in the other. Different alloys of the same nominal composition also showed differences in stability when compared with one another. By selecting a particular variety of the Chromel type alloy and matching it to another selection from the constantan samples, the compensation properties could be optimized for least drift.

Klapper and associates (1965) described halogen attack on a Chromel-constantan couple when the temperature exceeded the limits of a low temperature wiring insulation. It was recommended that this couple not be used at above 200°C where fluorine or chlorine might be present.

Sparks, Powell, and Hall (1968) gave calibration tables, curves, and thermoelectric power data for the 0-280 K temperature range. These workers (1970), and Hust, et al. (1971) recommended the couple for general engineering work above liquid hydrogen temperatures, because of reasonably good homogeneity as well as low thermal conductivity for both alloys. This gives a range from 20 K to about 900 or 1000°C for Chromel-constantan, making it a particularly attractive choice where both low and moderate temperatures are measured. Von Middendorff (1971) evaluated magnetic field effects on this type of thermocouple at cryogenic temperatures and found that for a field of 60 kG the maximum increase of thermoelectric power occurred at 25 K and was about 2.5 per cent.

Bianchi, et al. (1966) exposed thermocouple hot junctions to an integrated flux of 4×10^{20} n/cm^2 in a nuclear reactor. The maximum temperature was 650°C. No appreciable drift due to radiation damage was detected for couples which were precalibrated before installation and then remotely calibrated after irradiation. These thermocouples were recommended for nuclear application, within the integrated flux and temperature limits of the experiment.

Ni-14.7Cr vs Cu-40Ni. This combination was mentioned in a review article by Schulze (1939), and attributed to H. Lenz and F. Kofler. It was said to have been used at high temperature in the flames of coke oven and blast furnace gases.

Ni-(14-17)Cr-(6-10)Fe vs Cu-43Ni. A needle thermocouple of Inconel versus constantan was described by Rauch (1954). It was prepared and calibrated in a manner similar to that for Fe-28Ni-18Co vs Cu-43Ni, except that a 1.93 mm Inconel tube was used, within which the insulated constantan wire was inserted. The Inconel composition given here is the nominal composition associated with that alloy; other elements may also be present. The voltage-temperature curve is shown in Figure 4-3; the output is somewhat greater than that of copper-constantan. At 500°C the thermoelectric power is 68 μV/°C. At 599°C the output was reported to be 36.59 mV, and was stable during six consecutive heating-cooling cycles.

Ni-25Fe-11Cr vs Cu-45Ni. Chromel-X versus Copel was mentioned by Finch (1962) and was said to be useful for temperatures below 200°C. It was recommended by the manufacturer[7] of Chromel-P vs Alumel for this temperature condition.

Ni-16Mo vs Ni-50Cu. Schulze (1939) attributed the development of this thermocouple to W. Rohn, during the 1924-1927 period. It can be used in air to about 800°C without excessive oxidation. The output as shown by Schulze is linear, or nearly so, and is given in Figure 4-3. From the curve, the thermoelectric power is 86 μV/°C, even greater than that of Chromel-constantan.

Ni-(15-25)Cr vs (96-99)Cu-(0.5-3.0)Ni-(0-1.5)Mn. This combination may also have small optional amounts of manganese and carbon in the positive conductor. Zysk (1967) described these compositions for use as extension wires with the W-3Re vs W-25Re thermocouple. The temperature range is preferably between 0 and 200°C.

THERMOCOUPLES WITHOUT COPPER

Antimony, Bismuth, Germanium, and Tellurium Bearing Combinations

Sb vs Bi. Hunt (1964), in his review of the early history of the thermocouple, stated that Seebeck experimented with a number of materials, including antimony and bismuth, during his studies which led to the discovery of thermo-electricity in 1821. He found Sb vs Bi to be the most effective in terms of the deflection of a magnetic needle, which detected the electric current in the circuit.

The first thermocouples and thermopiles for thermal radiation detectors were of Sb vs Bi. Smith, Jones, and Chasmar (1957) described early developments in this field, beginning with work by Melloni in the 1830 period. His thermopiles were usually of Sb vs Bi and he was able to greatly extend observations in the infrared spectrum by their application. During the next half century this was the most widely used type of detector for infrared work. Interest then shifted to other methods of detection until relatively recent times.

Most modern applications of this and similar combinations involve thin film thermocouples and thermopiles. Cartright (1932, p. 221) prepared thin film evaporated metal couples of Sb vs Bi, and shortly thereafter Harris (1934) described multijunction radiometers deposited by sputtering the metals on thin cellulose. Other work at this time was contributed by Johnson and Harris (1933) and Jones (1934). Strong (1938, p. 305) described methods of preparing both vapor deposited and sputtered thermocouple junctions, as well as fine wire couples. Roess and Dacus (1945) deposited thin film couples on Formvar substrates for radiation detectors in infrared spectrographs. Harris (1946) developed a technique for evaporating the metals and obtaining relatively low resistances and high thermoelectric powers for the thin films. A radiation thermopile was described consisting of 50 junctions, with a receiving area of about 0.11 cm² and 70 ohms resistance. Watson (1966) included thin film couples in a review of surface temperature measurement methods, referring to work with Sb vs Bi by Allen (1954) and Bennett (1959). A high impedance, 57,000 ohm, 89 junction thermopile was described by Staffsudd and Stevens (1968), produced by an evaporative technique on a thin membrane. Usefulness was demonstrated for detection of wavelengths up to 330 μ for the far infrared.

Walters (1966) observed that the thermoelectric properties of vapor deposited antimony and bismuth films were highly dependent upon deposition rate and other factors, and concluded that bulk properties for both materials can be obtained by proper control of the deposition parameters. Koike and co-workers (1968) studied the effect of film thickness on antimony and bismuth in terms of junctions of each material with silver. It was found that bulk properties were exhibited for thicknesses of more than 10,000 Å. Earlier, Johnson and Harris (1933) had observed a size effect for bismuth films, but not for those of antimony. Gerhardt (1962) reported decreases of several per cent in the magnitude of the thermoelectric power after three months exposure to air of thin films of each thermoelement. Protective coating experiments were made in an effort to minimize this effect.

Most of the results of thin film work report a thermoelectric power S of 50 to 75 μV/°C. Roeser and Wensel (1941, p. 1293) gave outputs for bulk antimony and bismuth relative to platinum which would result in values of 12.23 mV at $100°C$ and 23.71 mV at $200°C$ for Sb vs Bi. The Handbook of Chemistry and Physics (1960) indicates that a bulk metals couple using commercial bismuth would have a value of S of 79.3 $\mu V/°C$ at $0°C$, while the use of electrolytic bismuth would give a value of $110 \mu V/°C$ at $0°C$. These figures are all higher than those usually given for thin film couples.

Pb vs Bi. This combination was mentioned by Hawkings (1964) in a review of thermocouples and neutron flux monitors for in-core reactor measurements. Lead and bismuth were used in a neutron thermopile, a device sensitive to the temperature rise caused by the heat of nuclear reactions in a stacked set of lead and bismuth discs. The work was attributed to F. S. Replogle and associates.

Te vs Bi. Schulze (1939) reviewed earlier results for specially constructed bulk metal thermocouples of these elements used to measure radiant heat. A thermoelectric power of about 360 $\mu V/°C$ was given. The high resistivity of tellurium is a disadvantage, and rapid drift occurs at temperatures above about $90°C$, also due to the tellurium. Work with these elements was also reported by Cartwright (1932, p. 73), followed by Harris and Johnson (1934), and Harris (1934), who prepared a thermocouple for a radiometer by sputtering the metals on thin cellulose. The thermoelectric power S was on the order of 400 $\mu V/°C$, but the resistance was 150,000 ohms. Gerhardt (1962) studied evaporated couples where thin membranes of Mylar and other materials were used as substrates. Separate measurements of S relative to copper for both tellurium and bismuth resulted in values of +332 to +335 $\mu V/°C$ for tellurium vs copper, and –40.8 to –58.7 $\mu V/°C$ for bismuth vs copper, for freshly prepared films of a number of different specimens. For seven Te vs Bi thin film couples which were prepared, the resistance ranged from 920 to 1400 ohms, the average being 1100 ohms.

Gerhardt observed aging effects after measuring freshly prepared films versus copper and then repeating the measurements three months later. Decreases in thermoelectric power relative to copper of 1 to 17 per cent were reported. Thin protective coatings were studied in an effort to minimize this effect.

Walters (1966) found that the thermoelectric properties of thin film Te vs Bi couples were affected by deposition rate and other parameters, as was also reported for Sb vs Bi. It was stated that bulk thermoelectric properties could be produced by proper deposition technique.

Te vs Ni-3Mn-2Al-1Si. Gerhardt (1962) included a study of thin film tellurium versus a deposited alloy produced from the evaporation of Alumel. The composition of the deposited film was not given, but Gerhardt was aware of the possibility of compositional changes during the deposition process.

Sb-48.3Cd vs Bi. A fine wire couple of this composition was described by Jones (1934). The wires were about 0.03 mm in diameter, prepared by the

Taylor process of drawing out the metals in glass capillaries, then etching the glass away.

Sb vs Bi-10Sb, Sb-25Cd vs Bi-10Sb, Sb-35Cd vs Bi-10Sb. These combinations were mentioned by Hornig and O'Keefe (1947) for use in fine wire thermopiles. The pair with antimony was suggested because of its high thermoelectric power S, given as 118 μV/$^{\circ}$C. The Sb-25Cd vs Bi-10Sb thermocouple was reported to have an even higher value of 190 μV/$^{\circ}$C for S. The couple with 35 per cent cadmium was slightly less sensitive, S being 184 μV/$^{\circ}$C. Bismuth-antimony alloys have also been considered as the negative conductor for cryogenic thermocouples because of the high thermoelectric power below 75 K (Anon., 1967).

Te-(0.4-0.5)Bi vs Bi-10Sb, Te-1.5S vs Bi-10Sb. Also mentioned by Hornig and O'Keefe (1947) for thermopile usage, the value of S for the former pair was given as 269 μV/$^{\circ}$C and that of the latter as about 650 μV/$^{\circ}$C. The electrical resistivity of Te-1.5S was high, however, and was given as 350,000 μohm cm.

Bi-5Sn vs Bi. Strong (1938, p. 311, 333) gave a detailed description of the construction and properties of fine wire thermopiles for radiometers using these materials. A figure of 120 μV/$^{\circ}$C was given for the thermoelectric power.

Bi-10Sn vs Bi. Details on the construction of a radiometer were described by Cartwright (1932, p. 221), but no thermocouple performance information was included.

Bi-5Sn vs Bi-3Sb. Hornig and O'Keefe (1947) chose this pair for use in a fine wire radiometer, based on the results of their analysis of radiation detector requirements. Thermocouples, and 2 and 4 junction thermopiles were prepared. Wire diameters were on the order of 0.02 mm. The thermoelectric power was 105 μV/$^{\circ}$C.

Sb_2Te_3-$x$$Bi_2Te_3$ vs Bi_2Te_3-$y$$Bi_2Se_3$. A short section on radiometer applications of these materials was included by Abowitz and associates (1963) in a report primarily concerned with direct energy conversion of heat to electrical power. Experiments were made with thin film thermocouples on glass slide substrates. A post-deposition annealing process at 300 to 350°C was required because of pronounced changes of film properties which occurred during the first heat exposure after deposition. A number of different film compositions were investigated and the thermoelectric power relative to a reference metal was determined, as well as other properties. Table 4-4 gives the compositions of a few of the samples tested, showing additional small amounts of materials that were added as dopants. The approximate thermoelectric powers for paired samples are also shown.

Ge(p) vs Ge(n). Doped thin film thermocouples of germanium vs silver and germanium vs germanium were prepared by Onuma (1965). Glass or mica sub-

strates were used. The thermoelectric power S of the Ge(p) vs Ge(n) thermocouple was 770 $\mu V/°C$ between 0 and 300°C.

TABLE 4-4

Nominal Composition and Thermoelectric Power of Selected Paired
Samples of Thin Film Thermocouple Materials

Composition	S $(\mu V/°C)^a$
Sb_2Te_3-25Bi_2Te_3-2.3Te vs Bi_2Te_3-25Bi_2Se_3-6CuBr	291
Sb_2Te_3-25Bi_2Te_3-2.3Te vs Bi_2Te_3-25Bi_2Se_3-0.1CuBr	320
Sb_2Te_3-25Bi_2Te_3-2.3Te vs Bi_2Te_3-25Bi_2Se_3-0.1AgI	321

[a]These are approximate values for the 25 to 125°C interval.

Tyushev and Shelud'ko (1969) described similar thin film Ge(p) vs Ge(n)
thermocouples deposited on ceramic substrates. Thicknesses ranged from 600 to
1500 Å; the value of S was 730 $\mu V/°C$ and the thermocouple resistances were in
the order of several thousand ohms. The voltage-temperature characteristic was
linear from 0 to 300°C. Detailed data on the electrical properties of the films
were included. Values for S of +310 $\mu V/°C$ and –420 $\mu V/°C$ were given for the
p and n type films, respectively.

Kendall, Dixon, and Schulte (1967), and Schulte and Kohl (1970) used
bulk single crystal doped germanium difference thermocouples for measurement
of surface heat flux in a wind tunnel model study. The p-type germanium was
gallium doped; the n-type material was antimony doped. These units had a
thermoelectric power of 2000 $\mu V/°C$ between 0 and 50°C.

Ge vs Si.[8]   Kendall and coworkers (1967) also used these materials in a prototype heat flux sensor, but no details were given.

Chromium, Iron, and Nickel-Bearing Thermocouples

Fe vs Al.   Studennikov and Erkovich (1965) included this pair in a study of
possible extension wire combinations for use with tungsten-rhenium alloy thermocouples. Unannealed iron wire and electrical aluminum wire were used. The
thermoelectric power was 15 $\mu V/°C$ in the 0-50°C range.

Fe vs Ni.   Wenzl and Morawe (1927) reported the use of an Fe vs Ni thermocouple for blast furnace temperature measurements up to about 1350°C. When
compared with a noble metal couple good agreement was observed.

A steel versus nickel thermocouple was used by Bendersky (1953) for
surface temperature measurements on the bore surface of gun barrels. The construction consisted of a nickel wire with a nickel oxide insulation layer, held
within a steel tube. The assembly was installed in the gun barrel and the end of

the wire-tube assembly was polished flush with the bore surface. A coating consisting of a 1-micron layer of nickel was used to establish the junction between the end of the wire and the tube. Earlier work of this type had been done by P. Hackemann in Germany during World War II. Moeller (1962, 1963) described similar applications.

Morrison and Lachenmayer (1963) described evaporative deposited thin film couples on glass substrates. The film thicknesses were normally 3000 Å and a coating of silicon monoxide was applied over the films for protection. Variations in the calibration curves were small, amounting to about 2°C. A number of advantages were given for using iron and nickel, including ease of evaporation, good adherence to the substrate, and the avoidance of alloys in the evaporative process, where changes in composition may occur during deposition. Marshall, Atlas, and Putner (1966) described similar silicon monoxide coated thin film couples. The glass slide substrate temperature was maintained at 250°C or more during deposition to avoid crazing, which otherwise resulted when film thickness exceeded 2000 Å. Each couple was calibrated between 20 and 300°C using a silicone oil bath. Experimental error during calibration was about ±2°C.

The calibration curves were found to be linear over the temperature range of 20 to 260°C. Values of the thermoelectric power S ranged from 20 to 27.5 $\mu V/°C$. These figures are lower than bulk metal values. A voltage-temperature curve for bulk metal Fe vs Ni, using data from Roeser and Wensel (1941, p. 1293) is shown in Figure 4-4. The value of S in the 100 to 200°C interval can be found from this curve to be about 33 $\mu V/°C$. Calibration drift reported by Marshall, et al., was within the experimental error of ±2°C during a six-week period when measurements were made.

Reed and Ripperger (1971) fabricated very small thermocouples by vapor depositing iron and nickel on 2-micron diameter quartz filaments. The calibration curve agreed with a conventional wire thermocouple calibration. A probe configuration with a 20-micron diameter was described and fabrication of smaller probes was considered to be possible. The small size of these thermocouples resulted in a very rapid response time for a step change of temperature in the surrounding medium.

Mo vs Ni.   A voltage-temperature curve from work by Roeser and Wensel (1941, p. 1293) is plotted in Figure 4-4. Potter (1954) made a calibration from 0 to about 1230°C, a portion of which is also shown in the figure, and also constructed a calibration table. The data were believed to be accurate within about ±1°C for the particular batches of molybdenum and nickel wires which were used. The agreement between the two curves in Figure 4-4 is good, considering that for thermocouples utilizing molybdenum small differences in the composition of wires from different sources may produce unusually large changes in thermoelectric characteristics. Thermoelectric power S, from Potter's data, is 35, 50, and 53 $\mu V/°C$ at 480, 1040, and 1220°C, respectively.

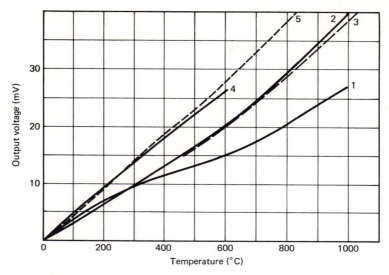

FIGURE 4-4 Thermocouples with nickel or nickel alloys, I. 1. Fe vs Ni, Roeser and Wensel (1941, p. 1293). 2. Mo vs Ni, Roeser and Wensel (1941, p. 1293). 3. Mo vs Ni, Potter (1954). 4. Ni-10Cr vs Ni, Rauch (1954). 5. Ni-18Mo vs Ni, General Electric Co. (undated).

Jensen, Klebanoff, and Haas (1964) found this combination superior to Pt-10Rh vs Pt for a condition where it was necessary to attach thermocouples directly to nickel surfaces. A total of 12 Mo vs Ni thermocouples were tested at 950°C in a vacuum, after aging for 30 minutes at about 1000°C. A drift of 0 to –1 per cent occurred during the first 3 days at 950°C. After 24 days the drift ranged from 0 to –4 per cent. Pt-10Rh vs Pt thermocouples which were similarly tested showed drift of as much as –30 per cent after 24 days.

Ni-10Cr vs Fe. The combination of Chromel versus iron was selected by Fanciullo (1964) for use as extension wires with W-3Re vs W-25Re for a special application. The output of Chromel-iron was 6 per cent less than that of the tungsten alloy couple at a temperature of 100°C. At somewhat higher temperatures this error decreased, becoming zero near 190°C.

Ni-(9-10)Cr vs Ni. Rohn (1924, 1927) studied the nickel-chromium alloys and gave this combination. Here the chromium content of 9 to 10 per cent results in the highest output that can be obtained with nickel-chromium alloys relative to nickel. At the same time, both metals have good oxidation resistance. From the curve given by Rohn, the output was nearly linear and was slightly above 40 mV at 1100°C.

Rauch (1954) calibrated a Chromel-nickel wire thermocouple by comparison with a Chromel-Alumel pair for possible application in a needle thermocouple. The voltage-temperature curve is given in Figure 4-4 and is quite similar to that of Chromel-Alumel. The thermoelectric power is 40 μV/°C in the 500°C region.

Ni-14.7Cr vs Ni.   Schulze (1939) described Chronin versus nickel in a review article and said that it had been studied for high temperature measurements in the flames of coke ovens and blast furnace gases, proving efficient up to 1100°C. Development work was attributed to H. Lenz and F. Kofler.

Ni-16Cr vs Ni.   Rohn (1927) suggested the use of this high chromium content alloy as a substitute for Ni-10Cr because of its improved oxidation resistance, which was three times better than that of nickel. The output of this couple is about 34 mV at 1000°C, compared with 37 mV for Ni-10Cr vs Ni.

Ni-16Cr-24Fe vs Ni-4.3Si.   This and a number of very similar combinations were studied by Ihnat (1959) and Ihnat and Hagel (1960) during the development of extension wires for the Pt-15Ir vs Pd thermocouple. The nominal positive alloy composition given was that of Nichrome. The calibration curve was quite close to the curve for Pt-15Ir vs Pd.

Ni-10Cr vs Zr.   Shoens and Shortall (1953) included this combination during the evaluation of Zr vs Cu-43Ni and Zr vs Ni-3Mn-2Al-1Si, both of which were considered preferable to Chromel vs zirconium because of higher thermoelectric powers. A calibration by means of the boiling point method was described and the results are shown in Figure 4-5. The thermoelectric power was 16.63 μV/°C in the 100-200°C range.

Ni-34Fe vs Ni.   According to Schulze (1939), this couple can be used up to 1000°C. The alloy was specified as nickel steel, and considerable variation in performance was observed, depending on the exact alloy composition. The calibration curve appears in Figure 4-5. The thermoelectric power was given as 21.7 μV/°C between 300 and 400°C, and 32.0 μV/°C between 700 and 800°C.

Ni-18Mo vs Ni.   Lever (1959) and Ihnat (1959) mentioned this thermocouple. A calibration table by the developer[9] shows an output of 67.80 mV at 1310°C, the highest temperature given. A curve is plotted from the table in Figure 4-4 for temperatures up to about 830°C. The thermoelectric power is 56 μV/°C at 800°C. According to the manual on temperature measurement of the American Society for Testing and Materials (1970, p. 48) thermocouples now available[10] are slightly different in composition from that given above, the negative alloy in particular being composed of nickel with about 1 per cent cobalt. The couple can be used in hydrogen or other reducing atmospheres, inert atmospheres, or vacuum, up to at least 1200°C. The metals oxidize too readily for use above about 650°C in an oxidizing atmosphere.

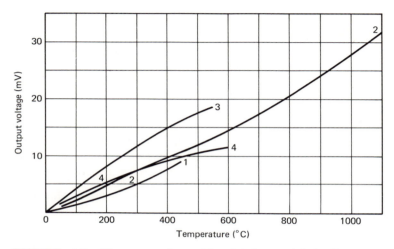

**FIGURE 4-5   Thermocouples with nickel or nickel alloys, II.
1. Ni-10Cr vs Zr, Shoens and Shortall (1953).   2. Ni-34Fe vs Ni,
Schulze (1939).   3. Mo vs Fe-29Ni-17Co, Cox (1963).   4. Zr vs
Ni-3Mn-2Al-1Si, Shoens and Shortall (1953).**

Bundy (1961) investigated the effect of high pressure by tests up to 72 kB, while a 100°C temperature difference was present between the hot and the cold junctions. An additive pressure correction was found which compensated for thermocouple pressure sensitivity. The correction was of the same order of magnitude as that for Chromel-Alumel.

Mo vs Fe-29Ni-17Co.[11]   Molybdenum-Kovar was investigated by Cox (1963) for use in high vacuum apparatus. Several thermocouples were calibrated by comparison with Chromel-Alumel. An equation was developed for the 50-550°C range, the data from which were within 0.5 per cent of the calibration data for couples made from five different melts of molybdenum and from eight of Kovar. The voltage-temperature curve is given in Figure 4-5. The thermoelectric power S is 40.8 μV/°C and 24.4 μV/°C at 100 and 500°C, respectively.

Changes in S with temperatures near 0 and 550°C were discussed. The former was a rather abrupt change said to be the result of the γ to α phase change in Kovar. The latter was presumed to be caused by behavior of the alloy near its Curie point.

Zr vs Ni-3Mn-2Al-1Si.   Shoens and Shortall (1953) evaluated thermocouples of zirconium with other metals for temperature measurements of zirconium-clad fuel elements in a nuclear reactor, where it was desired to disturb the coolant flow and neutron flux as little as possible with extraneous materials. Zirconium versus Alumel and Zr vs Cu-43Ni were selected as being the most worthy of consideration.

Calibrations were made between 100 and 450°C by the boiling point method, using water and four other substances. An equation was developed from the data and from this a table was prepared for voltage vs temperature in the 0-600°C range. A voltage-temperature curve is given in Figure 4-5 using figures from the table. At 100 and 500°C, the thermoelectric power S is 25.5 and 12.0 μV/°C, respectively. The value of S, particularly at higher temperatures, is considerably less than that of Zr vs Cu-43Ni. The errors in calibration were higher for zirconium-Alumel, primarily because of the lower thermoelectric power, which led to an accuracy estimate of ±5°C for the calibration of this thermocouple.

Co-20Cr vs Co-5Fe.   This was suggested by Wagner and Stewart (1962) for applications up to about 927°C. The thermoelectric power is 41.5 μV/°C and 37.6 μV/°C at 600 and 800°C, respectively.

Co-23.5Fe vs Co-17Fe.   Villamayor (1968) developed this combination after a study of binary cobalt-iron alloys. It is useful as an extension wire pair for W-5Re vs W-26Re for temperatures up to 250°C at the extension junctions with the tungsten alloys. With a 250°C temperature at that location, and 2000°C at the tungsten-rhenium alloy thermocouple hot junction, the extension wire error was 1 per cent.

(95-97.5)Ni-(1.5-2.2)Cr-(0-1.2)W-(0.7-1.1)Al-(0.3-1.0)Si vs Ni-(3-5)W.   Metals with compositions in this range were specified by Zysk and Osovitz (1970) for use as extension wires with the W-3Re vs W-25Re thermocouple.

Ni-10Cr vs Ni-xMn-yAl-zSi.[12,13]   A base metal thermocouple with the trade name of Chromel-P vs Alumel was developed by the Hoskins Manufacturing Company and was in use for industrial applications during the World War I period. Adams (1920) reviewed the thermocouples in use at that time and included tables and other data on this couple. Lohr (1920) outlined its development in terms of the alloy selection process which led to a choice of the metals now used. Reference tables were prepared by Roeser, Dahl, and Gowens (1935) at the National Bureau of Standards, and later by Shenker and coworkers (1955) in NBS Circular No. 561. Jones (1969) converted the NBS tables to the International Practical Temperature Scale of 1968 for the range of –180 to +1300°C in one degree increments. Gottlieb (1971) gave data for correcting IPTS-48 tables to IPTS-68.

Chromel-P vs Alumel has been supplemented by similar couples produced by other manufacturers. At present these thermocouples match the NBS 561 reference tables within ±2.2°C from 0 to 277°C, and ±3/4 per cent from 277 to 1260°C; all are designated as Type K thermocouples, using the classification system accepted by most manufacturers and interested organizations. Common usage of the original name has persisted, nonetheless, so that Chromel-Alumel is often used as a synonym for the Type K designation. Special alloys with improved performance characteristics have also been developed; these may or may not meet Type K specifications depending upon the particular alloys used. Some of these thermocouples and their nominal compositions are listed[14] in Table 4-5.

TABLE 4-5

Chromel-P vs Alumel and Similar Thermocouple Alloys

Type	Trade Name	Nominal Composition	Ref.
K	Chromel-P vs Alumel[a]	Ni-10Cr vs Ni-3Mn-2Al-1Si	e
K	Kanthal-P vs Kanthal-N[b]	Ni-10Cr vs Ni-(2-3)Si	f
K	Tophel vs Nial[c, d]	Ni-9.3Cr-0.55Fe-0.4Si vs Ni-2.5Mn-2Al-1.05Si-0.5Co-0.25Fe	f
–	Chromel-P + niobium vs Special Alumel[a]	Ni-9.41Cr-0.35Si-0.2Fe-0.2Nb vs Ni-1.23Si-0.26Fe	g

[a] Hoskins Mfg. Co.
[b] The Kanthal Corporation.
[c] Wilbur B. Driver Co.
[d] Elements present in amounts of less than 0.1 per cent are not shown.
[e] Finch (1962).
[f] Caldwell (1962).
[g] Potts and McElroy (1962).

According to Starr and Wang (1969), the recommended range for Type K thermocouples is generally listed as 0-1260°C with the tolerances as given earlier; premium or special grade wires are available for the same range with tolerance limits of one half of the standard tolerance. A voltage-temperature curve is plotted in Figure 4-6, using data from the tables of Shenker, et al. (1955). The thermoelectric power changes by small amounts over the range from 0 to 1000°C but is always near 41 $\mu V/°C$ in this range.

Chromel-Alumel thermocouples can be used in the cryogenic range. Powell, Caywood, and Bunch (1962) developed a cryogenic calibration table for 0-280 K, and Sparks, Powell, and Hall (1968) gave similar information. Both of these sources included tabulated thermoelectric power data, and the latter reference included curves of thermoelectric power versus temperature.

Powell and associates also showed homogeneity results for Type K and other commonly used thermocouples. Type K couples were somewhat less homogeneous than Cu vs Cu-43Ni couples. Also, below about 60 K the thermoelectric power S of the Type K couple was shown to be less than that of Cu vs Cu-43Ni. At 10 and 20 K, S is 1.98 and 4.14 $\mu V/K$, respectively. Cryogenic use of this thermocouple is limited at temperatures below about 20 K, compared with such combinations as Ag-0.37at.Au vs Au-0.07at.Fe and Ni-10Cr vs Au-0.07at.Fe.

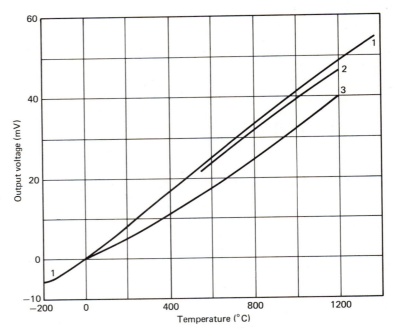

FIGURE 4-6   Chromel-Alumel and other thermocouples.   1. Ni-10Cr vs Ni-3Mn-2Al-1Si, Shenker, et al. (1955).   2. Ni-9.5Cr-1.5Si vs Ni-4.25Si-0.25Mg, Burley (1971).   3. Ni-20Cr-1Nb-1Si vs Ni-3Si, Guettel (1956).

Roeser and Wensel (1941, p. 284) reported that there had been little success in attempting to fit equations of the form

$$V = aT + bT^2 + cT^3$$

to the calibration curve in the 0 to 300°C interval. An equation based upon three calibration points at about 100, 200, and 300°C led to errors of up to 1°C. A better fit was described for the –190 to 0°C range using three about equally spaced points. The resulting equation was in error by not more than 2 μV at any point.

Chromel-Alumel type thermocouples cannot always be calibrated and used with as high an accuracy and repeatability as some of the other thermocouples, even though they are often clearly superior in other respects when the slightly greater errors are acceptable. In some applications these relatively small inaccuracies can be removed by heat treatment prior to calibration, or by not exceeding a temperature limit.

Lack of repeatability and its consequences was studied by a number of investigators, and their results showed that a region of reversible metallurgical instability exists in Chromel type alloys between about 300 and 550°C. Within this region the thermoelectric power depends upon the temperature, the exposure time, and the prior temperature-time history of the thermocouple. Callcut (1965) suggested that a short range crystal lattice ordering effect is responsible. Burley (1970, 1971) assembled considerable supporting evidence, including his work, that by Fenton (1969), and numerous other results from the literature. Fenton (1971) described a number of observations resulting from studies by means of the traveling gradient method of evaluating inhomogeneities.

In brief, a Chromel type wire at room temperature that has been freshly hard drawn, heavily cold worked, or quenched from a high temperature state is in a fully disordered condition. However, this is a metastable state and at higher temperatures a detectable drift of thermoelectric power takes place over a period of time as short range ordering occurs. According to Burley (1971), it takes about 40 weeks for a stable condition to be reached when the wire temperature is 300°C. At 350°C only 4 days is required; at 400°C 1.25 hours is required. At 450°C a one-minute exposure results in complete short range ordering, and above this temperature ordering is virtually instantaneous, until the order-disorder transition temperature is reached. The upper limit for the instability region depends on this temperature, since the wire is always in a stable disordered state above the transition. The effects of the reversible instability region on thermocouple performance are discussed in following paragraphs.

If an as-drawn Type K thermocouple is exposed to a slowly increasing temperature and is heated from 0°C to a temperature a little above 500°C, and is then slowly cooled to its original value, it exhibits a hysteresis-like voltage-temperature loop. This is caused by changes of thermoelectric power in the instability region. Had the couple been heated above a temperature of about 550°C and then slowly cooled, the voltage loop would be somewhat different in appearance, because the instability process in that portion of the wire at a temperature above 550°C would reverse and that section of wire would revert to its original condition until cooled through the instability region. On the other hand, for heating only to about 300°C no instability is encountered, although some information suggests the possibility of long term drift in the neighborhood of 250°C and above when periods of many weeks are considered. It should be noted that in many practical cases where other error-producing factors predominate or where only short term use is intended it may be quite feasible to operate up to 350°C without observing the small instabilities discussed here.

In the stable region the limiting factor on repeatability was said by Fenton (1967) to be the inhomogeneity of the wire, which generally leads to errors of about ±0.25°C. This is considerably better than the errors that may result from exposure to temperatures in the instability region, although these are usually less than 2°C. At worst an 8°C apparent temperature deviation may result, according to Burley (1971).

The above description applies to an as-drawn thermocouple, which is initially work hardened and in a disordered state. By annealing the entire wire length at a temperature in the instability region, the wire becomes stable for use up to the order-disorder transition temperature. Presumably, the Ni-10Cr has reached its maximum degree of short range order. As before, repeatability is limited by inhomogeneities and is normally about ±0.25°C.

Various investigators have given somewhat different results for the maximum use temperature that can be achieved with heat treated couples without onset of instability caused by disordering at high temperature. Fenton (1967) gave a figure of 460°C; Sibley, et al. (1968) stated a limit of 540°C. Burley (1970) showed successful results of stabilizing heat treatment at 550°C followed by calibration to 550-570°C.

Sibley, Spooner, and Hall (1968) considered the effects of the instability condition at higher temperatures, as well as the 0-540°C range, since Type K couples are often used at temperatures well above 540°C. The changes which occur in the 300-540°C range were referred to as aging, although it must be kept in mind that it is a reversible type of aging. Wire which is allowed to age in the 300 to 540°C range cannot be used at higher temperatures without the occurrence of instability as the wire reverts to its original condition. Therefore, for full range use of Chromel-Alumel from 0 to 1260°C, some reversible instability may be encountered whenever the hot junction temperature exceeds 300°C. This is true even when the hot junction is above the upper instability limit because some portions of the wires are still in the 300-540°C region; portions of the length of Chromel type wire then tend toward different thermoelectric properties, resulting in inhomogeneities.

Sibley and coworkers indicated that for 0 to 1260°C use deviations of about 3°C or less may result for thermocouples which are used repeatedly over the full temperature range. These are considerably smaller deviations than Burley's figure of 8°C, but the temperature cycling conditions are different for the two cases. Plots of temperature-cycled thermocouples given by Green and Hunt (1962) show some of the changes which may be encountered. Deviations can be decreased to some degree by avoiding changes of thermocouple immersion depth and by calibrating in place where possible.

Another small error effect was observed by Wintle and Salt (1967) during studies of a Chromel-Alumel type couple which met British Standard 1827 within tolerances of ±3°C to 400°C and ±3/4 per cent above 400°C. It was found that couples made from the same batch of wires generally exhibited differences of ±0.25 per cent of the indicated temperature. Also, the calibration curves deviated sharply, although by small amounts, from what would normally be called a smooth curve through a short temperature interval. Therefore a true difference curve could not be drawn without first plotting a large number of calibration points over the range of interest. An error limit tolerance of ±0.75°C was suggested for calibration curves with data points 20°C apart. In view of this condition it was also suggested that the specified tolerance for a given batch of wire should be not less than ±(0.75°C + 0.25 per cent of temperature) from the calibration obtained from one sample of that batch.

Type K thermocouples are commercially available as bare wires or with insulation, the latter type being obtainable in a wide variety of materials for particular applications. Also available are swaged or drawn thermocouple cables; the choice of sheath material and insulation must be considered from the standpoint of thermal expansion characteristics, particularly for long term use under thermal cycling conditions. Babbe (1964), Ferry (1964), and Lagedrost (1964) discussed this type of problem for Chromel-Alumel thermocouples. Carr (1971) described improved methods of determining the degree of insulation compaction in sheathed cables.

Commercial thermocouple wires may be cold worked to some degree as received from the manufacturer, depending upon the supplier and the order specifications. Preparation for use may cause additional amounts of cold work, especially when considerable handling is required for difficult installations. These conditions lead to errors, particularly those caused by inhomogeneities which result when wires are bent or kinked. Reduction of cross sectional area by cold working an entire length of wire also has an effect on thermocouple output.

Lever (1959), Potts and McElroy (1960, 1961, 1962), and Pollack and Finch (1962) investigated cold working of various types for Type K wires. Chromel was found to be much more affected than Alumel. Usually cold work resulted in changes of less than 10°C in thermocouple calibration. Proper heat treatment reduced cold working effects by nearly 90 per cent in the case of Chromel and somewhat less for Alumel; this resulted in thermocouples which were stable to within ±0.5°C at 400°C. Results by Fenton (1971) included the finding that recovery from cold work takes place in Alumel at room temperature. The effects of a 25 per cent elongation were just discernible after six month's storage.

Adamenko (1969) described cold working tests where an 11 per cent deformation was produced with specimens of Alumel, Chromel, Copel, and constantan. Of the various combinations obtainable from these metals, Chromel-

Alumel was the least sensitive to a percentage change of thermoelectric power caused by plastic deformation. Lister and Harvey (1967) described cold working of Chromel-Alumel during handling, assembly, and testing of tensile specimens. It was found that rigid control of these procedures was required to maintain accuracy.

As a general rule, Type K thermocouples can be used in air over long periods of time, and for short periods where there is a deficiency of oxygen.[15] In such marginally oxidizing atmospheres the Chromel wire may develop a greenish coating, often referred to as green rot. Short term operation can be conducted in vacuum or inert atmospheres and this can be extended to lengthy periods of time in very clean systems where no oxygen is present. Preferential vaporization of chromium from the positive conductor at high temperatures then becomes the limiting factor for long term usage, because of the resulting thermocouple drift as the chromium content decreases. According to Campari and Garribba (1971) this is a relatively minor effect. Operation in reducing atmospheres is not recommended because of the rapid drift which results. In particular, drift occurs in atmospheres containing hydrogen, sulphur, or carbon monoxide. Operation is possible, nevertheless, in atmospheres of pure dry hydrogen or cracked ammonia. According to the manual of the American Society for Testing and Materials (1970, p. 21), usage in these atmospheres is acceptable provided that the dewpoint is below $-40°C$.

Dahl (1941, p. 1238) reported on the first comprehensive drift tests of bare wire Chromel-Alumel as well as other base metal thermocouples. Some of the results are given in Table 4-6, which indicates that long exposure to an atmosphere of clean air results in an increase of thermocouple output and therefore an apparent increase of temperature.

Drift tests were also conducted with wires of 3.26, 1.02, and 0.64 mm diameter to investigate the effect of wire size on drift. For 1000 hours at 650°C, drift was nearly the same for all three wire sizes and did not exceed 1.4°C. For 1000 hours at 870°C there were still no pronounced increases of drift depending on wire size. At the highest calibration temperature of 870°C the drift ranged from about 1.5°C for the 3.26 mm wire to slightly over 4°C for the 0.64 mm diameter wire. Higher drifts occurred in the 200-310°C calibration region where changes of up to 5°C or 1.5 per cent were shown. Wire oxidation was clearly evident and there was a measurable decrease in diameter of wires exposed at the higher temperatures. The Alumel wire showed the greater drift tendency at high calibration and exposure temperatures. At the higher temperatures where oxidation was measurable, the diameter of the Alumel wire decreased faster than that of the Chromel. More recent drift test data by Burley (1971) correlated quite closely with the results of Dahl.

Gatward (1941) compared positive drift observed in oxidizing atmospheres with negative drift, which is often seen in measurement work. The latter was described as due to the presence of reducing gases. When traces of such gases are present the resulting contamination may lead to decreased positive drift or to negative drift, depending on the amounts present.

TABLE 4-6
Drift of Chromel-Alumel Thermocouples in an Atmosphere of Air

Exposure Temperature (°C)	Exposure Time (hours)	Wire Size (mm)	Maximum Apparent Change of Temperature (°C)
427	1000	1.02	< 0.6
538	1000	1.02	< 0.6
649	1000	3.26	+ 1.1
760	1000	3.26	1.7
871	1000	3.26	2.8
982	1000	3.26	4.4
1093	1000	3.26	10.6
1204	200	3.26	11.7

Tests by Potts and McElroy (1962) and Dobovišek and Rosina (1968) showed considerably higher drift rates than those of Dahl, for air exposure at comparable temperatures. Potts and McElroy's results, however, were for wires that had been exposed at a uniform high temperature over the entire length; Dahl's experiments were conducted with the thermocouple hot junction at high temperature and portions of the wires in a temperature gradient leading to much cooler temperatures. The test conditions and procedures of Potts and McElroy were probably much more severe than those of Dahl, which would explain the differing results. In the case of Dobovišek and Rosina the resulting drift was in the direction of decreasing apparent temperature rather than the positive drift usually associated with exposure in clean air. From the information given it was not clear as to why this condition was experienced.

Pumphrey (1947) observed the progressive oxidation of both Chromel and Alumel wires under purely oxidizing conditions at temperatures near 800°C. Alumel was affected more than Chromel, as Dahl had reported earlier, and exhibited a more pronounced brittle surface layer. Sibley (1960) described the development of a special Alumel alloy with improved oxidation resistance compared with the regular alloy. Potts and McElroy (1960, 1961, 1962) described a special Alumel, as well as other binary nickel alloys with silicon. The composition of the special Alumel was given as Ni-1.23Si-0.26Fe. Chromel plus added niobium, Ni-9.41Cr-0.35Si-0.2Fe-0.2Nb, was also suggested for added oxidation resistance. The combination of Chromel-P + niobium versus a special Alumel would have improved drift and oxidation characteristics compared with the Type K thermocouple.

Hughes and Burley (1962) made 3000-hour drift tests and longer period metallurgical investigations of Chromel and Alumel specimens obtained from a number of different sources. It was shown that Alumel type samples with high silicon content and low manganese and aluminum contents were more drift and oxidation resistant than the other Alumel specimens or the Chromel samples. The nominal compositions of the best Alumel type alloys were Ni-1.36Si-0.38Mn, and Ni-2.30Si-0.32Mn. Both samples contained traces of a number of other elements. The superior performance of these samples was attributed to a barrier effect caused by silica or silicate films at the metal-oxide interface, and from the virtual absence of readily oxidizable alloying additions.

Burley (1969) determined that Ni-10Cr, exposed to air at high temperatures, experienced a pronounced depletion of the chromium by conversion to chromium oxide in the region near the wire surface, during the oxidation process. The core remained initially undisturbed, but with increased exposure time the depth of the depletion region increased. A reasonable correlation between computed and experimental thermocouple drift was shown, based on relating the depletion of the chromium to the change of the thermoelectric voltage. The oxidation mechanism in Alumel was said to be relatively complex compared with that for Chromel.

Burley's work showed that chromium near the metal surface in Ni-10Cr alloys oxidizes faster than the nickel when the atmosphere is air. This condition becomes much more pronounced when less oxygen is available. Marginally oxidizing atmospheres, inert atmospheres with traces of oxygen, or vacuum with oxygen traces from out-gassing or leakage all produce preferential or selective oxidation of chromium.

Bennett, Rainey, and McClain (1962) and Rainey and Bennett (1963) found that the presence of an oxide layer on Type K thermocouple wires supplied sufficient oxygen to preferentially oxidize chromium in the wires when an otherwise inert atmosphere was present. Their results showed that a negative thermocouple drift resulted. The oxidation process occurred along the portion of the wire that was in the 800 to 900°C temperature range. Therefore, clean, bright-finished wires should be used with atmospheres other than air. It is also desirable to avoid inert atmospheres with traces of oxygen, or air with a deficiency of oxygen to prevent the same type of condition.

Spooner and Thomas (1955) discussed the decrease of thermocouple output in clean protection tubes with large length-to-diameter ratios, where lack of air circulation resulted in a deficiency of oxygen. The cause of the voltage decrease was found to be the result of the remaining oxygen preferentially attacking the chromium. The condition could be eliminated by increasing the supply of available air through the use of a larger diameter protection tube, or by a flow of air along the wire. As an alternative it was shown that titanium metal could be used as a getter to scavenge oxygen from a closed tube and permit oxygen-free operation.

Sibley (1960) described a special Chromel-P which was much less drift sensitive to preferential oxidation than regular Chromel-P wires. Obrowski and Von Seelen (1961) described similar results with other nickel alloys containing chromium and small amounts of additional elements. Bennett, Rainey, and McClain (1962) suggested the use of a Ni-20Cr alloy, which was considered to be less susceptible to preferential oxidation. The manual on thermocouple usage by the American Society for Testing and Materials (1970, p. 44) described commercially available thermocouples with generally improved performance compared with Type K couples. These are Thermo-Kanthal special,[16] Tophel II-Nial II,[17] and Chromel 3-G-345 vs Alumel 3-G-196.[18] Other thermocouples with nickel-chromium versus high nickel content alloys, which show improved performance under various conditions, are separately described in later sections of this chapter.

Pumphrey (1947) may have been the first to describe consequences of preferential oxidation of chromium from Chromel, although that particular explanation was not known at the time. Pumphrey did explain another condition, which consisted of thermocouple behavior in still air with the presence of contaminants. It was found that embrittlement associated with progressive oxidation in air above 800°C is considerably increased by the presence of oil or other carbonaceous or sulphur-bearing matter. This was particularly the case for Alumel. Wires and protection tubes should be clean to avoid such materials and their effects.

Direct contact of bare wires with other metals can also be a source of error or of thermocouple failure. Cormier and Claisse (1967) discussed the effects of contact with titanium and zirconium, observed by them and by others. In one case the thermocouple wire, which had been spot welded in place, detached when the metals alloyed and then melted at the attachment point.

In atmospheres other than air, McCoy (1966) investigated the behavior of thermocouples in carbon monoxide, carbon dioxide, and mixtures of the two. In carbon monoxide a 0.81 mm diameter couple showed little drift for 200 hours at 820°C, but then began to drift rapidly downward in output. Both thermocouple wires were found to contain over 5 weight per cent carbon near the hot junction after the test.

Drift was also evident for 0.55 mm diameter thermocouples in a mixture of carbon dioxide with 5 per cent by volume of carbon monoxide. At 650°C an initial positive calibration change of 7°C was observed, followed by additional slow upward drift of about 3°C during 2500 hours exposure. A similar test conducted at 750°C resulted in an initial upward calibration change of 16°C followed by a further slow drift to +19°C at 80 hours exposure. Then the direction of drift reversed and by the end of a 2500-hour total exposure the indicated temperature was 50°C lower than the true temperature. A third test was conducted at 850°C and positive instability was not observed. The initial drift was negative, and was about –4°C after 20 hours. At the end of 2500-hours exposure the drift indicated an apparent temperature decrease of about $-120°C$. After tests of this type, the thermocouple wires were very brittle, and heavy oxidation was observed.

Tests of 257 hours duration in carbon dioxide with 0.81 mm diameter wires showed negative drift. Drifts of -9, -8, and $-80°C$ were observed for temperatures of 620, 722, and $833°C$, respectively. Wire embrittlement was not mentioned, but other tests of 1000 hours duration did result in embrittlement. Somewhat similar tests at $850°C$ by Dobovišek and Rosina (1968) with 1 mm diameter wires also resulted in negative thermocouple drift. The drift was -4.50 mV, measured at $900°C$ after 90 hours, and was equivalent to about $-110°C$.

The latter investigators also conducted exposure tests in sulphur dioxide at $850°C$. Negative drift was rapid and large, reaching $-4°C$ after 1 hour and nearly $-100°C$ after 4 hours, for a calibration temperature of $900°C$. Caldwell (1962) indicated that Alumel is embrittled by sulphur to the point of breakage. Sibley (1960) found that the special oxidation-resistant Alumel showed no improvement in resistance to sulphur.

Dubovišek and Rosina reported tests in a reducing atmosphere from a bed of charcoal and stream of nitrogen at a temperature of $850°C$. After 90 hours a thermocouple showed a negative drift of $-65.3°C$ at the calibration temperature of $900°C$, a decrease of slightly over 7 per cent. A decrease of 2.7 per cent was observed after only 6 hours exposure. Early work by Hougen and Miller (1923) had shown sensitivity to reducing atmospheres, there being no drift after 50 hours at $900°C$ but a -4 per cent drift after an additional 50 hours at $1000°C$. Silica protection tubes were used in this case, however, and were partially effective at $900-1000°C$ for at least some of the exposure period. Ford (1951) patented a protective coating of chromic oxide for use on the wires, which was said to prevent the reaction of reducing atmospheres with the thermocouple.

Sheathed thermocouples are not immune to contamination and drift. Rainey, Bennett, and Hemphill (1963) concluded that graphite-contaminated atmospheres, hydrogen, and carbon monoxide could affect stainless steel sheathed thermocouples. However, Long (1966) reported that by copper plating the stainless steel sheath, attack of the sheath by graphite-contaminated helium atmospheres at $750°C$ could be prevented. Metallographic examination revealed no change of the Type K thermocouple wires; no voltage drift data were given, however.

The behavior of stainless steel sheathed thermocouples with insulated hot junctions was investigated by Evans and Holt (1967). The insulation used was magnesium oxide, and tests were conducted in an atmosphere of carbon dioxide. At $400°C$ drift occurred and was essentially complete by the end of 100 days exposure, with maximum apparent temperature changes of $+4°C$ but more normally about $+2°C$. This behavior, as well as numerous other results of Evans and Holt and of Fenton (1967), suggest that the reversible instability condition described earlier was usually the predominant factor affecting performance. This condition and the methods for minimizing it are probably even more important with respect to sheathed couples than with bare wires, since other disturbing factors are usually less prominent in the former case.

Design, fabrication, and long term drift tests of sheathed thermocouples were discussed by Fanciullo (1964, 1965), Bliss and Fanciullo (1965), and Bliss (1971). It was shown that for lengthy operation of sheathed thermocouples at high temperature, the cool temperature end of the sheath must be sealed to prevent atmospheric contamination. The most advanced cable design consisted of aluminum oxide as insulation, with tantalum-foil-lined Nb-1Zr tubing. Insulated thermocouple junctions were used. It was necessary to maintain rigid quality control of the insulation purity and fabrication methods in order to eliminate drift-producing impurities from the cables. Tests of 10,000 hours in helium at about $1100°C$ resulted in a maximum drift of -1.45 per cent. Post-test examination indicated relatively minor metallurgical changes in the test specimens.

Williams and coworkers (1968) used thermocouples in magnesium oxide insulation with various types of sheaths. These were embedded in a graphite block and exposed to an atmosphere of argon at $980°C$ in the presence of tellurium. Type 347 stainless steel was found to be the most satisfactory sheath for this application. The enclosed thermocouple began to drift downward in apparent temperature after 140 hours. Couples in Type 310 stainless steel and Hastalloy X sheaths did not perform as well.

Type K thermocouples have been used extensively in nuclear applications because of their low sensitivity to transmutation-produced contamination. Kelly (1960), Kelly, Johnston, and Baumann (1962), Bliss and Fanciullo (1966), Bianchi, et al. (1966), Gottschlich and Bellezza (1966), Martin and Gabbard (1971), and Obrowski, et al. (1971) reported nuclear radiation exposures for total integrated flux ranging up to about 6×10^{20} n/cm^2 without significant drift due to transmutation of thermocouple materials. Hesse (1969) reviewed the general use of sheathed Chromel-Alumel type thermocouples for reactor applications. Aarset and Kjaerheim (1971) described experiences in measuring coolant, fuel cladding, fuel surface, and fuel center temperatures. Flemons and Lane (1971) reported on the use of sheathed thermocouples, including their design, methods of installation, and performance, for surface temperature measurements in nuclear fuel bundles. Liermann and Tarassenko (1971) described instrumentation for in-pile measurements as well as analytical and experimental studies of sheathed thermocouple cable behavior both in and out of pile.

Carpenter, et al. (1971) tested tantalum-sheathed couples with magnesium oxide insulation for 4470 hours at about $975°C$. The integrated thermal flux was 1.6×10^{21} n/cm^2. The integrated fast flux for energies greater than 0.18 meV was 2.7×10^{21} n/cm^2. There was no significant drift observed when brief in situ calibrations at $1083°C$ were made during the test.

Hoitink, et al. (1970) exposed Chromel-Alumel thermocouples to an integrated flux of 7×10^{21} n/cm^2 at a temperature of 566°C. Irradiation effects were investigated by calibration before and after exposure. Changes of calibration approaching 1 per cent were observed after irradiation, and deviations occurred during the post-irradiation calibration. The latter changes suggested that crystalline structural damage by fast neutrons had occurred which was then removed by annealing during the calibration process. Transmutation effects may also have been present.

Dau, Bourassa, and Keeton (1968) reviewed earlier studies of short term reversible effects on calibration by Boorman (1950), Andrew, et al. (1952), Sturm and Jones (1954), and work at the Oak Ridge Graphite Reactor (Anon., 1956), as well as later results. It was concluded from these and other data for different types of thermocouples that nuclear dose rate effects are small for flux of less than 10^{13} n/cm^2 sec. Worsham and Warren (1969) reported errors of less than 0.1°C for Chromel-Alumel in a flux of 4×10^{13} n/cm^2 sec.

Leonard (1969), in a discussion of work by several investigators, concluded that there was some conflict in the earlier results as to magnitude and time dependence of calibration changes, if any, that can be expected at radiation levels commonly encountered in reactors. It was shown that the relative locations of the flux and the temperature gradients along the thermocouple wires have a major influence on calibration changes. It was found that for in-pile thermocouple installations, temperature gradients should be restricted to locations outside high flux regions, in order to minimize possible radiation-dependent calibration changes.

Type K thermocouples are also used for temperature measurements in high pressure cells. According to Bundy (1961), the first work with this couple was done by F. Birch in 1939, but no significant pressure effect on thermocouple voltage was measured with the equipment of that period. Bundy found that pressure did affect the thermocouple. Other results were reported by Hanneman and Strong (1965, 1966), Peters and Ryan (1966), and Getting and Kennedy (1970). Studies of performance in a combined environment of high pressure and high temperature are difficult, and there are some discrepancies among the results of the various investigators.

Getting and Kennedy found that at moderate pressures and temperatures the thermocouple indicated a temperature equal to or only slightly less than the true temperature. This behavior has led to use of the Type K thermocouple as a reference for comparison of other combinations which are more sensitive to pressure in this range. At higher pressures and temperatures, however, the indicated temperature was found to be higher than the true temperature, and the required correction for temperature was comparable in magnitude with that for Pt-10Rh vs Pt under the same conditions.

Correction curves were given, based on experimental work to 35 kB and 1000°C, and extrapolated to 50 kB and 1200°C. From the curves, at 35 kB and

1000°C the correction is –510 μV or about –12.5 Celsius degrees, since the indicated temperature is too high. The uncertainty of the correction was said to be ±(20% +20 μV) for the experimental portions of the curves, which leads to an uncertainty of ±0.2 per cent in the measured temperature, after correction. The latter figure is small because the correction is small. A detailed discussion of the use of the correction curves as well as of the experimental work itself was included.

Chromel-Alumel type thermocouples show some sensitivity to the presence of magnetic fields, due primarily to the magnetovoltage generated in the Alumel. Loscoe and Mette (1962) found that the effect is small except in intense fields. Alumel is the most sensitive of the commonly used base metal thermocouple materials in this respect. This places a possible limitation on the use of Type K thermocouples when fields of kilogauss magnitude are present.

Ni-9.5Cr-1.5Si vs Ni-4.25Si-0.25Mg.   This thermocouple was first described by Burley (1971) and referred to as Nicrosil I vs Nisil. The voltage-temperature curve in Figure 4-6 is slightly lower than the curve for Chromel vs Alumel. The thermoelectric power is about 37 μV/°C at 1000°C. The addition of silicon to both alloys greatly increased their resistance to oxidation. Magnesium also aided this condition as used in the negative alloy. Drift tests demonstrated the improvement which, compared with Chromel-Alumel type thermocouples, amounted to an order of magnitude or more in some cases. The behavior caused by reversible instability was said to be comparable with the behavior of Type K couples caused by that condition.

The drift tests were conducted with thermocouples in air. The wire diameters were 0.81, 3.26, and 6.54 mm. From the drift versus time curves shown, for a 1000°C exposure temperature there was a positive drift of about 1°C after 100 hours, but very little additional drift over a total period of slightly more than 1400 hours. The curve for 1100°C showed small positive and negative changes in equivalent temperature that were within a few tenths of a degree during 700 hours exposure. At 1200°C drift was negative and within 1°C in magnitude during about 670 hours.

Similar test results were shown where thermal cycling was conducted at several times during the exposure period. Such a cycle consisted of cooling to 400°C in 15 hours and reheating to 1200°C in 5 hours. Nicrosil I vs Nisil was affected very little by this procedure, but Type K couples showed substantial output variations. Other tests where the temperature gradients along the wires were changed while maintaining a constant hot junction temperature showed that this thermocouple was much less affected by inhomogeneities than Type K units, following lengthy high temperature exposure.

Ni-15Cr-1.5Si vs Ni-4.25Si-0.25Mg.   Burley (1971) proposed Nicrosil II vs Nisil for good oxidation resistance plus freedom from the reversible instability associated with Type K and Nicrosil I vs Nisil thermocouples. A considerable amount of evidence favoring these properties was described. Experimental work had not yet been started at the time of the report.

Ni-14Cr vs Ni-3Si.   Zysk and Robertson (1971) reviewed a communication from C. L. Guettel and his associates that described the use of Ni-14Cr for improved performance compared with Ni-10Cr. By using the Geminol N alloy[19] as the negative conductor, the resulting thermocouple showed improved characteristics compared with the Type K couple. The voltage-temperature curve was somewhat different from the Type K characteristic, however. This work was done as early as 1961.

Ni-22Fe-15Cr vs Ni-4.3Si.   Ihnat (1959) and Ihnat and Hagel (1960) described the use of Nichrome[19] versus Ni-4.3Si as extension wires for the Pt-15Ir vs Pd thermocouple. The nominal composition shown here was given by Jaffe and Hallinan (1961) for the same application. According to Ihnat and Hagel, the optimum match for use with Pt-15Ir vs Pd was obtained when the Nichrome was annealed for 30 minutes at 1000°C and the Ni-4.3Si was annealed for 1 hour at about 950°C. The deviation between the extension wire calibration and the thermocouple calibration was within about 0.7 per cent up to 700°C. The extension wire output was 1.4 per cent lower than that of Pt-15Ir vs Pd at 784°C. Exposure of an extension wire thermocouple at 1000°C for 528 hours resulted in a drift of +0.42 per cent.

Ni-20Cr-1Nb-1Si vs Ni-3Si.[20]   This thermocouple was described by Guettel (1956) and was developed for use in industrial reducing atmospheres where Ni-10Cr vs Ni-3Mn-2Al-1Si and similar thermocouples with about 10 per cent chromium content deteriorate. The output is lower than that of Chromel-Alumel, but at higher temperatures the sensitivities of the two types are quite comparable. The voltage-temperature curve is shown in Figure 4-6. Caldwell (1962) gave values of thermoelectric power of 32.2, 38.9, and 40.0 μV/°C at 427, 871, and 1204°C, respectively.

The nominal composition above is commercially supplied[21] under the trade name of Geminol P versus Geminol N; the temperature range was given by Caldwell as −18 to +1260°C. Wires are available that meet tolerances of ±2°C up to 260°C and ±0.75 per cent from 260 to 1260°C, with respect to the manufacturer's curve.

The positive Geminol alloy is affected by the same type of reversible metallurgical instability that occurs in Ni-10Cr and which was discussed in detail under Ni-10Cr vs Ni-xMn-yAl-zSi. The range of instability for Geminol is somewhat different, but the resulting temperature measurement errors are small, as with Chromel-Alumel type couples. Callcutt (1965) suggested that the instability is caused by short range ordering. Burley (1970) showed that after a 2-hour stabilization treatment at 600°C Geminol can be calibrated and used up to 650°C in a stable condition. Without stabilization, hysteresis-like calibration loops resulted with maximum temperature differences of about 7°C between ascending and descending portions of a temperature calibration cycle. Similar results were obtained for Chromel-Alumel type couples. Typical instability errors with the latter are about 3°C or less during practical applications; therefore similar performance can be expected of Geminol thermocouples.

Guettel (1956) described tests conducted in a reducing atmosphere of flowing nitrogen with 16 per cent H_2, 10 per cent CO, 5 per cent CO_2, and 1 per cent CH_4. A 954°C temperature was chosen, which produced a condition known to accelerate green rot attack on nickel-chromium alloys in a reducing atmosphere. After a 212-hour exposure, a Geminol thermocouple drifted by the equivalent of a 3.3°C temperature increase. A shorter test over a 171-hour period resulted in a positive drift of about 0.3°C. Thermocouples with lower chromium content show negative drifts with many times the magnitudes of these values under similar conditions. From these results, it is evident that Geminol thermocouples can be used in reducing atmospheres similar to that of the tests; it should be noted that no sulphur was present, however.

Potts and McElroy (1960, 1961, 1962) showed that both Geminol alloys had superior oxidation resistance characteristics. Rainey and Bennett (1963) indicated that Geminol P is less susceptible to marginally oxidizing atmospheres than Ni-10Cr, and is not affected by preferential oxidation. Geminol P did not show the negative drift which is characteristic of Ni-10Cr under these conditions. Northover and Hitchcock (1968) found that the Geminol thermocouple alloys showed good stability during short period heating between 50 and 750°C, from the as-received condition. Evans and Holt (1967) exposed stainless steel sheathed thermocouples with magnesium oxide insulation to an atmosphere of carbon dioxide. Drift was less than that of Chromel-Alumel type thermocouples under similar conditions.

Helm (1967) stated that Geminol thermocouples gave satisfactory performance in contact with graphite for temperatures up to 800°C. A total of 45 of these thermocouples were then used in a nuclear radiation environment with neutron exposures which varied from 0.5×10^{21} to 2×10^{21} n/cm^2, for energies exceeding 0.18 meV. Temperatures ranged from 400 to 800°C. In general, a downward shift of about 4 per cent in output over this range was observed after irradiation.

Ni-14.7Cr vs Ni-27.1Fe-11.2Cr-2Mn.   This combination, called Chronin versus Cekas, was mentioned by Schulze (1939) and attributed to H. Lenz and F. Kofler, who studied it for work in flames from coke ovens and blast furnace gases up to 1100°C.

Ni-19.7Cr-1.0Si-0.8Ti-0.5Mn-0.3Fe-0.01C vs Ni-3Mn-2Al-1Si.   Bertodo (1963) reported the development of this combination for use as extension wires with the W-2(ThO$_2$) vs SiC (nitrogen-doped) thermocouple. Within a temperature range of 0 to 250°C the voltage deviation between the extension wires and the thermo-couple was 5°C or less in magnitude.

Fe-(17-19)Cr-(8-10)Ni-(0-2)Mn-(0-1)Si vs Ni-3Mn-2Al-1Si.[22]   This is Type 302 stainless steel versus Alumel, and was considered by Popper and Zeren (1966) during a search for extension wires for W-3Re vs W-25Re. The characteristics were very similar to those of a Type 308 stainless steel versus Alumel thermo-couple, which was preferred by the investigators.

Fe-(18-20)Cr-(8-12)Ni(0-2)Mn-(0-1)Si vs Ni-3Mn-2Al-1Si.[22]   Type 304 stainless steel versus Alumel couples were used by Chand and Rosson (1965) for the mea-surement of heat flux in a Type 304 stainless steel plate at temperatures in the 20-100°C range. The design was similar to that of Bendersky (1953), and Moore and Mesler (1961).

Fe-(19-21)Cr-(10-12)Ni-(0-2)Mn-(0-1)Si vs Ni-3Mn-2Al-1Si.[22]   Popper and Zer-en (1966), and Popper and Knox (1967) described Type 308 stainless steel versus Alumel for use as extension wires with the W-3Re vs W-25Re thermocouple. With the extension wire connections in the 50 to 650°C range and the thermo-couple hot junction at 1500°C, the error limits due to the extension wires were –0.4 and +2.9 per cent. The magnitude of the percentage error decreased at higher hot junction temperatures.

Fe-(33.5-34.5)Ni-(19.5-20.5)Cr-(0-0.15)C vs Ni-(4.25-4.75)W.   Metals of these compositions were described by Zysk, et al. (1969) for use as extension wires with the Pt-40Rh vs Pt-3Rh thermocouple. These extensions were said to match the thermocouple output of the noble metal couple quite closely over the 0 to 1000°C range. The extension wires were stable at temperatures up to about 850°C in an oxidizing atmosphere and to about 1000°C in an inert atmosphere.

THERMOELECTRIC DATA ON MISCELLANEOUS BASE METALS

The voltage-temperature curves for a number of base metals versus a reference metal of platinum are given in Figures 4-7 through 4-15. Little or no additional information was given in the literature sources for many of these materials, and in other cases the individual metals rather than thermocouple pairs were the subject of the studies.

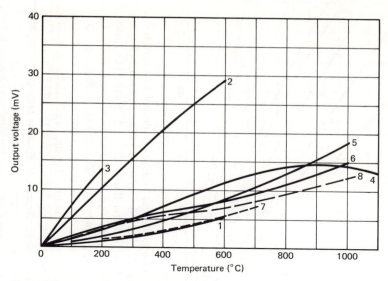

FIGURE 4-7   Thermoelectric voltage of selected base metals relative to platinum, I.   1. Al vs Pt;   2. Sb vs Pt;   3. Pt vs Bi;
4. Pt vs Co;   5. Cu vs Pt;   6. Fe vs Pt: Roeser and Wensel
(1941, p. 1293).   7. Mg vs Pt, Rohn (1927).   8. Pt vs Ni, Roeser
and Wensel (1941, p. 1293).

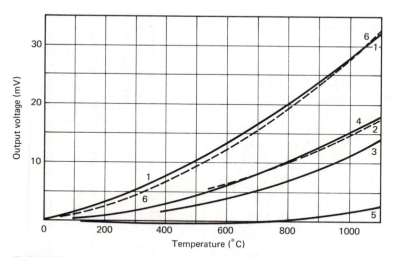

FIGURE 4-8   Thermoelectric voltage of selected base metals relative to platinum, II.   1. Mo vs Pt, Roeser and Wensel (1941, p.
1293).   2. Nb vs Pt, Fanciullo (1964).   3. Re vs Pt, Sims and
Jaffee (1956).   4. Ta vs Pt;   5. Th vs Pt;   6. W vs Pt: Roeser
and Wensel (1941, p. 1293).

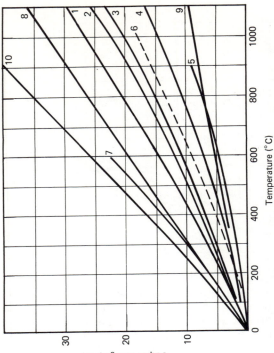

FIGURE 4-9 Thermoelectric voltage of selected base metals relative to platinum, III. 1. Ni-10Mo-1Mn vs Pt; 2. Ni-20Mo-1Mn vs Pt; 3. Ni-9.4Cr-5Mo vs Pt; 4. Fe-1Mo-1Mn vs Pt; 5. Fe-7Mo-1Mn vs Pt; 6. Ni-19.1Cr-4.8Si vs Pt: Hunter and Jones (1941). 7. Ni-3.85Cr-2.11 Mn-0.11Si vs Pt (Alloy: 105, Driver-Harris Co.); 8. Pt vs Ni-0.28Mn--0.17Cr-0.03Si; 9. Pt vs Ni-0.08Si-0.02Mn (Alloy: A Nickel, International Nickel Company): Lohr, Hopkins, and Andrews (1941). (Reference junction temperature is 25°C for curves 1 through 6, and 20°C for curves 7 through 9.)

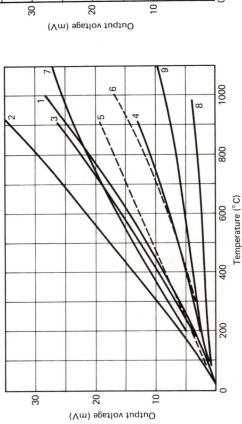

FIGURE 4-10 Thermoelectric voltage of selected base metals relative to platinum, IV. 1. Ni-13.15Cr-5.58Fe-1.02Mn-0.24Si-0.05C vs Pt (Alloy: Nirex, Driver-Harris Co.) 2. Ni-18.85Cr-1.00Mn-0.69Si-0.33Fe-0.03C vs Pt (Alloy: Nichrome V, Driver-Harris Co.) 3. Ni-20.59Fe-14.93Cr-2.02Mn-0.35Si vs Pt (Alloy: Nichrome, Driver-Harris Co.) 4. Fe-36.40Ni-16.02Cr-1.37Mn-0.14Si vs Pt (Alloy: 525, Driver-Harris Co.): Lohr, Hopkins and Andrews (1941). 5. Fe-18Cr-8Ni vs Pt; 6. Ni-24Fe-16Cr vs Pt; 7. Pt vs Cu-25Ni; 8. Ni-10Cr vs Pt (Alloy: Chromel-P, Hoskins Mfg. Co.); 9. Pt vs Ni-3Mn-2Al-1Si (Alloy: Alumel, Hoskins Mfg. Co.); 10. Pt vs Cu-43Ni (Alloy: constantan Reference junction temperature is 20 C for curves 1 through 4. The composition was not specified and may vary considerably from the nominal composition shown here): Roeser and Wensel (1941, p. 1293).

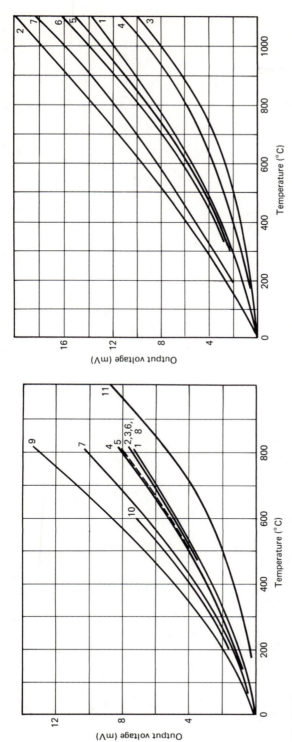

FIGURE 4-12 Thermoelectric voltage of selected base metals relative to platinum, VI. 1. Co-10Cr vs Pt; 2. Co-20Cr vs Pt; 3. Co-38Mn-4Al vs Pt; 4. Co-25Mn-10Cr-5Al vs Pt; 5. Co-25.5Mn-15Cr vs Pt; 6. Co-35Mn-10.2Cr-4.8Mo vs Pt; 7. Co-20Cr-2Ta vs Pt: Wagner and Stewart (1962).

FIGURE 4-11 Thermoelectric voltage of selected base metals relative to platinum, V. (Curves 1 through 9 are based on 3-point calibrations given in tabular form by Ihnat) 1. Type 302 Stainless Steel vs Pt; 2. Type 304 Stainless Steel vs Pt; 3. Type 305 Stainless Steel vs Pt; 4. Type 310 Stainless Steel vs Pt; 5. Type 316 Stainless Steel vs Pt; 6. Type 321 Stainless Steel vs Pt; 7. Type 330 Stainless Steel vs Pt; 8. Type 347 Stainless Steel vs Pt; 9. Ni-17Mo-15Cr-6Fe-5W-1Mn-1Si-0.15C vs Pt (Alloy: Hestelloy C, Haynes-Stellite Co.): Ihnat (1959). 10. Cu-4Sn-1Zn vs Pt, Roeser and Wensel (1941, p. 1293). 11. Co-10Mn vs Pt, Wagner and Stewart (1962).

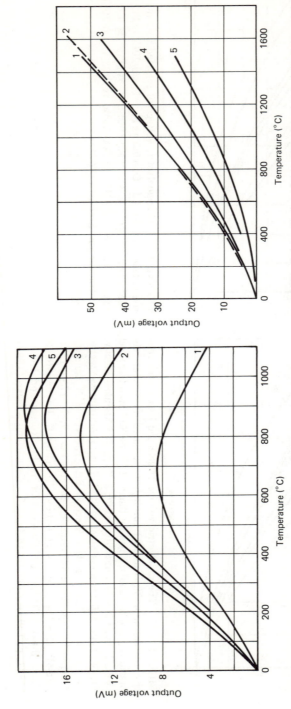

FIGURE 4-14 Thermoelectric voltage of selected base metals relative to platinum, VIII. 1. W-3Re vs Pt (doped W-3Re wire); 2. W-5Re vs Pt; 3. W-10Re-1(Th0$_2$) vs Pt; 4. W-15Re-2(Th0$_2$) vs Pt; 5. W-25Re vs Pt: Zysk and Robertson (1971).

FIGURE 4-13 Thermoelectric voltage of selected base metals relative to platinum, VII. 1. Pt vs Co-6Al; 2. Pt vs Co-40Ni; 3. Pt vs Co-25Ni; 4. Pt vs Co-30Fe; 5. Pt vs Co-15Fe: Wagner and Stewart (1962).

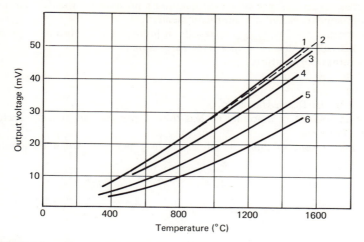

FIGURE 4-15   Thermoelectric voltage of selected base metals relative to platinum, IX.   1. Mo vs Pt;   2. Mo-5.77Re vs Pt;   3. Mo-10Re vs Pt;   4. Mo-20Re vs Pt;   5. Mo-30Re vs Pt;   6. Mo-50Re vs Pt: Zysk and Robertson (1971).

FOOTNOTES

1. All references are located at the end of the book.
2. Several alloys of high aluminum content were used.
3. Other elements that may be present in amounts of less than 2 per cent are not shown in this nominal composition.
4. The composition shown was that given by Rauch. Somewhat different compositions are sometimes given.
5. The name Manganin is also associated with compositions of lower copper content and with other elements added.
6. Somewhat different compositions have been given elsewhere.
7. Hoskins Mfg. Co.
8. Polarity was not specified.
9. The General Electric Co.
10. 20 Alloy vs 19 Alloy, manufactured by the Wilbur B. Driver Co.
11. Several other elements may be present up to specified limits of 0.5 per cent or less.
12. Alloys discussed in this section are basically nickel alloys with about 9 or 10 per cent chromium, paired with high nickel content alloys containing additional elements.

13. Type K thermocouples are included here.
14. The presence or absence of a manufacturer's product is intended neither as a recommendation nor as a nonrecommendation.
15. Overexposure to this condition results in preferential oxidation of chromium. This will be discussed in a later paragraph.
16. The Kanthal Corporation.
17. Wilbur B. Driver Co.
18. Hoskins Mfg. Co.
19. Product of the Driver-Harris Co.
20. Elements present in amounts of less than 1 per cent are not shown here.
21. Driver-Harris Co.
22. Elements present in amounts of less than 1 per cent are not shown for these alloys.

5

Base Metal and Nonmetal Thermocouples for Moderate and High Temperatures

REFRACTORY METALS

Thermoelectric Properties

The most important metals for high temperature thermocouples are tungsten, rhenium, and some of the tungsten-rhenium alloys. Molybdenum has been used to some degree, and a number of other elements are found in special thermocouple applications. Some general properties of these metals are described below.

Tungsten.   This element has a very high melting point of approximately 3410°C, and special fabrication methods are required to prepare wire, rod, or sheet stock because of the refractory properties of the metal. The element has been widely used for thermocouple work, both alone and alloyed with other elements. These metals must not be exposed to oxidizing atmospheres at high temperatures or to otherwise inert atmospheres containing small amounts of oxidizing impurities. Atmospheres of the inert gases or hydrogen are acceptable, and this is also true of high vacuum. In general, exposure to carbon is not recommended, although some exceptional conditions will be discussed under specific thermocouples.

   The tungsten wire which is available at present has a high degree of homogeneity and uniformity and is quite comparable with copper in this respect. The wire is normally available in the ductile state and can be easily bent and otherwise handled for installation. After the first high temperature heat treatment it is very brittle and breaks easily. Small changes of thermoelectric power also occur during the initial heating.

Since heat treatment results in embrittlement and changes in thermoelectric properties, it is best practice to install the wire, heat treat the thermocouples to a temperature above the highest application temperature, and then calibrate. Unfortunately this is not always practical, so a compromise must then be made in terms of whether an initially ductile state or high accuracy is the preferred condition. In many applications the relatively small change in thermoelectric power may be unimportant.

When tungsten is alloyed with some of the other elements embrittlement becomes less noticeable. This is true in the case of rhenium, but a relatively high percentage of rhenium is required in order to eliminate the embrittlement problem. Therefore the all-alloy tungsten-rhenium thermocouples are still affected to some degree by this condition, since one conductor must have a low rhenium content to provide a reasonable thermoelectric power relative to the other.

Improvement in ductility has been obtained by a wire manufacturing process where a tungsten-3 per cent rhenium mixture is doped with various volatile additives during the initial phase of wire fabrication by the powder metallurgy method. During high temperature heating these volatiles are driven off, but not before a favorable alteration of the alloy microstructure has resulted as a result of their presence. Some of the tungsten and W-5Re wires used for thermocouples have also been prepared in this manner. In another approach to the problem, tungsten alloys with other elements such as molybdenum have been used in place of tungsten-rhenium, with better ductility remaining after heat treatment.

Rhenium.   This metal has a melting point of 3180°C and is unlike tungsten because of its property of remaining ductile after exposure to high temperature. It work hardens rapidly at lower temperatures, however. It can be used in vacuum, inert atmospheres, or hydrogen, but not in oxidizing atmospheres. Unalloyed rhenium is satisfactory for thermocouple use to 2200°C, but is of limited value above this temperature because of a lack of homogeneity and uniformity in the wire. The nonuniformities make it impossible to establish a calibration table above 2200°C without excessive errors. There may also be large errors with individually calibrated couples because of the various effects associated with inhomogeneity. Tungsten alloys with up to 26 to 28 per cent rhenium are not similarly affected, however. Rhenium alloys with a few per cent tungsten have not been evaluated in this respect, although some thermocouple work has been reported.

Molybdenum.   This element has been used for some years as a thermocouple material. It has the property of becoming embrittled at high temperatures, but not to the extent that is typical of tungsten. The metal is extremely sensitive to impurities; these may change the thermoelectric power quite drastically. Because of this property it is also very sensitive to contamination. For consistent results thermocouples with molybdenum must be used in a clean system with vacuum, a highly purified inert atmosphere, or pure hydrogen. Different users have seldom agreed on the same calibration curve for such thermocouples as W vs Mo, and in some cases significant deviations are reported between batches of wire, or even between different thermocouples from the same batch.

Alloys of molybdenum with tungsten and other metals have been found to have good mechanical properties. In some cases results of thermocouple applications have been encouraging. However, at present there is insufficient information to make any general statements about thermoelectric properties and applications.

Niobium.[1] It has been found that niobium reacts with most vapors at elevated temperatures except for the inert gases, absorbs hydrogen, and may absorb at least some of the inert gases. Use in clean high vacuum systems or in argon may be acceptable.

Tantalum. This metal is even more affected by contaminants than molybdenum and is extremely sensitive to residual gases in vacuum systems. It absorbs hydrogen, carbon monoxide, and carbon dioxide. It is not recommended for thermocouple use, although some special applications have been described. Tantalum alloys may also be unsuitable for thermocouple use. Maykuth, Ernst, and Ogden (1965)[2] found this to be true for binary alloys of tantalum with molybdenum, niobium, titanium, and tungsten.

CERAMICS FOR THERMOCOUPLE INSULATION

The limitations of ceramic insulation materials become important for thermocouple work above about 1600 to 1800°C. The thermocouple materials and the ceramic may begin to react chemically, the electrical resistivity of the ceramic always decreases, and some insulators melt or undergo changes in crystal structure at temperatures in this region.

The presence of additional materials complicates both the determination of insulation performance and the design of high temperature sensors. Insulation impurities and progressive contamination are major factors in this situation. Until recently high purity materials were not commercially available and were difficult to prepare in the laboratory. As a result much of the information on materials behavior which is available at present is affected and is questionable. Lack of agreement between various investigators is not surprising. Some of the most recent work has made use of better quality insulation, however, and purity levels of 99.8 to 99.9 per cent have been reported.

A brief listing of insulation materials used with thermocouples is given in Table 5-1, which was prepared primarily from information by Anderson and

Bliss (1971), Fleischner (1971), and Zysk and Robertson (1971). Additional data from some of the reports given at the end of this section were also included. From these results, above 2000°C beryllium oxide, hafnium dioxide, and thorium dioxide appear to be the most suitable insulators for thermocouples, but even these have disadvantages.

TABLE 5-1
Summary of Insulation Materials for Thermocouple Applications

Material	Melting Temperature (°C)	Resistivity (ohm cm)[a]	Reactions with Metals[b] (at temperatures shown in °C)		Comments
			Mo	W	
Al_2O_3	2030	$>10^9$	2000	>1900	
BeO	2570	$>10^9$	1900	2100	Highly toxic in crushed or powdered form.
CaO	2600	10^4			
CeO	>2600	2×10^3, (800°C)			Oxidizing atmospheres only.
HfO_2	2840	2×10^5		>1500	
MgO	2800	2×10^7	1800	2000	Hygroscopic. Low resistivity above 1800-2000°C.
SiO_2	1700	5×10^4, (1077°C)	1500	1600	
ThO_2	3225	3×10^4	1900	2200	Slightly radioactive. High thermal neutron cross section.
Y_2O_3	2410	$\sim 10^7$			Low resistivity above 1800-2000°C.
ZrO_2	>2400	10^5	2150	2100	Low resistivity above 1800-2000°C.
BN		10^6 (850°C)			Vacuum to about 1800°C, nitrogen to about 3000°C.

[a] At 1000°C unless otherwise shown, as estimated from various sources. Disagreements of 2 to 4 orders of magnitude in reported data are common.
[b] In general these are temperatures for slight amounts of reaction.

A number of references give information on insulator physical properties, reactivity with metals, high temperature chemistry, and applications for thermocouple design. A brief summary of information was included by Caldwell (1962). A chapter with much useful data was given by Kohl (1967, p. 61). Other applicable information was included in work by MacKenzie and Scadron (1962), McGurty and Kuhlman (1962), Liddiard and Heath (1962), Hall and Spooner (1963, 1964), Clark (1964), Kamenetskii and Gul'ko (1965), Kuhlman (1966), Sessions (1966), Novak and Asamoto (1966, 1968), Asamoto and Novak (1967), Popper and Knox (1967), Meservey (1967), Jeffs (1967), Droege, et al. (1968), Fries, et al. (1969), Briggs (1971), and Trogolo (1971). The 1971 dated references given earlier are also applicable here.

THERMOCOUPLES WITH TUNGSTEN AND TUNGSTEN ALLOYS

Tungsten as One Conductor

W-5Mo vs W, W vs W-15 Mo.   Both of these pairs were investigated by Lachman and McGurty (1962, pp. 153, 177), who made calibrations in argon, hydrogen, and mixtures of the two gases. Curves for these couples are shown in Figure 5-1, and it can be seen that the outputs are low and the nonlinearities are extreme.

W vs W-25Mo.   This thermocouple is generally attributed to Pirani and Wangenheim (1925), who used it at temperatures up to 3000°C, but reported an unexpected temperature deviation at temperatures above 2700°C. A voltage-temperature curve from Schulze (1939) is included in Figure 5-1. From the curve, the thermoelectric power is about 4.1 μV/°C at 2000-2200°C.

W vs W-35Mo.   Lachman and McGurty (1962, pp. 153, 177) gave a calibration curve which is shown in Figure 5-1. The characteristics are similar to those of the W vs W-50Mo thermocouple.

W vs W-50Mo.   An investigation was made by Morgan and Danforth (1950) which resulted in an average calibration curve and table, using data for wires of different sizes and from different spools. The calibration work was done in a vacuum of 10^{-5} torr, and both the melting point method and optical pyrometry were used in establishing temperatures. A curve from these data appears in Figure 5-1; the thermoelectric power in the 1800-2000°C region is about 6 μV/°C. Lachman and McGurty (1962, pp. 153, 177) also gave calibration data, taken in atmospheres of argon, hydrogen, and mixtures of the two gases, during studies of the tungsten-molybdenum alloy system.

W vs Mo.   Fink (1912, 1913) used tungsten versus molybdenum for high temperature furnace measurements in an atmosphere of hydrogen, during an investigation of the properties of ductile tungsten wire which had become available during that period. Other investigators reported results shortly afterwards, but the first specific mention of work at temperatures above 2000°C was made by Pirani and Wangenheim (1925). A representative list of later calibration results shows the extent of the interest in W vs Mo: Morugina (1926), Van Liempt (1929), Osann and Schroeder (1933), Wilhelm, et al. (1948), Morgan and Danforth (1950), McQuillan (Undated; 1950), Greenaway, et al. (1951), McGeary (1952), Simons, et al. (1953), Nadler and Kempter (1961), Wong and Schaffer (1961), Lachman and McGurty (1962, pp. 153, 177), Hensel and Göhler (1964), and Pletenetskii and Mandrich (1965). A typical calibration, by Morgan and Danforth, is given in Figure 5-2. The thermoelectric power is about 7 μV/°C near 1600°C.

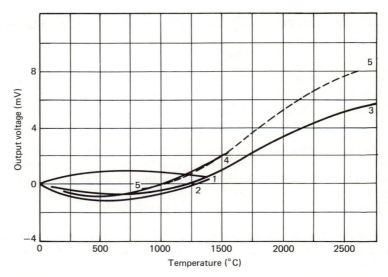

FIGURE 5-1   Thermocouples of tungsten and tungsten-molybdenum alloys.   1. W-5Mo vs W, Lachman and McGurty (1962, p. 153). 2. W vs W-15Mo, Lachman and McGurty (1962, p. 153).   3. W vs W-25Mo, Schulze (1939).   4. W vs W-35Mo, Lachman and McGurty (1962, p. 153).   5. W vs W-50Mo, Morgan and Danforth (1950).

When this combination is considered for use the fact must be faced that there are wide differences in the calibration results obtained by various investigators. The sensitivity of molybdenum to impurities and contamination is responsible for this condition. Therefore it is necessary to calibrate each thermocouple separately unless prior calibrations have established that a given batch of molybdenum wire is uniform. Some users have reported that calibration of one sample per batch is sufficient to establish a voltage-temperature curve. Others have found differences within a given batch. There is certainly every reason to suspect that the thermoelectric properties from batch to batch and from different sources of supply may vary considerably.

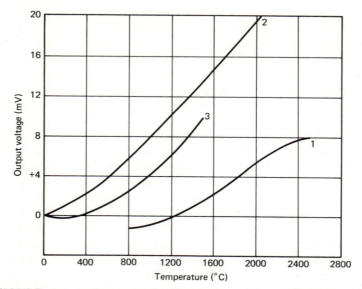

FIGURE 5-2   Thermocouples of tungsten with molybdenum and molybdenum alloys.   1. W vs Mo, Morgan and Danforth (1950). 2. W vs Mo-0.5Al, Adabaschjan (1959).   3. W vs Mo-1.0Fe, Schulze (1939).

Danishevskii (1959) described procedures for minimizing calibration variations by sorting and matching different wire pairs. Ergardt (1968) indicated that wire available in the Soviet Union is of sufficient uniformity that a standard thermocouple table has been adopted for applications work. The lack of reproducibility was mentioned however; also, a table taken from the standard, GOST 3044-61, gave millivolt values to only two significant figures, apparently because of the matching problem.

The temperature range is somewhat restricted by the reversal of polarity near 1300°C and declining thermoelectric power above 2400°C. There is some lack of agreement concerning the temperature of polarity reversal and there may be a maximum in the voltage-temperature curve at or above 2500°C, for some wire batches at least. Repeatability figures of ±5°C above 1700°C, and ±3°C below 1700°C, within a nominal range of 1300 to 2400°C, are typical of most results where such data have been included. The thermocouple can, of course, be used at lower temperatures providing that the reversal of polarity is acceptable, but a voltage minimum exists in the 600°C region with a value of between –1 and –2 mV depending on the particular calibration curve. This is an additional limitation to consider.

Simons, et al. (1953) reported that there were no significant calibration differences between untreated and previously heat treated specimens for the 0 to 1900°C range in either argon or nitrogen. Steven (1956), on the other hand, as well as McQuillan (1950), recommended the use of heat treatment. Greenaway and associates (1951), and Wong and Schaffer (1961), used heat treatment procedures. Treatment consisting of a few minutes exposure, preferably above the highest application temperature, appears to be desirable, although the resulting wire embrittlement may impede handling for calibration and installation. Where possible all handling should be completed before heating of any kind.

The W vs Mo thermocouple can be used in at least some reducing atmospheres, such as hydrogen. Operation in vacuum or argon is permissible, and also in nitrogen, to at least 1900°C. The atmosphere should be as impurity free as possible because of the sensitivity of molybdenum to contamination. Both elements oxidize too readily at high temperatures for operation of bare wire couples in air or other oxidizing atmospheres.

Calibration and use in a helium atmosphere is another topic where some disagreement exists. Troy and Steven (1950) performed a helium atmosphere calibration and several other investigators used that atmosphere for preliminary heat treatment. Pletenetskii and Mandrich (1965), however, quoted earlier work by A. A. Rudnitskii stating that use of helium resulted in instability, and then performed a test that demonstrated a change in the thermoelectric properties of molybdenum when exposed to helium with 0.1 per cent hydrogen. The result is still not clear-cut, nevertheless, because the reference metal with which molybdenum was paired was not identified; therefore, the question about contamination of the molybdenum from the reference metal, or vice versa, cannot be discussed beyond stating that the investigators had evidently ruled out that factor as a potential cause of the condition.

Potter and Grant (1949) reported that carbon, silicon, beryllium, and oxygen act as contaminants of the W vs Mo couple. Simons, et al. (1953) stated that instability produced by the presence of graphite resulted in a 25°C increase of apparent temperature after 15 minutes exposure at 1700°C. McGeary (1952) found no deterioration in the presence of zirconium or hafnium.

Thermocouple insulation usage also requires some discussion. Potter and Grant indicated that stabilized zirconium dioxide was compatible with W vs Mo and that beryllium oxide protection tubes appeared to be satisfactory. Greenaway, et al. (1951), and McQuillan (1950) reported instances of beryllium oxide attack, but Troy and Steven (1950) were of the opinion that such attack was caused by inadequately fired beryllium oxide insulators which contained occluded oxygen. Greenaway and coworkers also reported that magnesium oxide insulation was usable to at least 1900°C without a marked effect on the thermocouple, although some intergranular contamination of the thermocouple bead was noted. No contamination by aluminum oxide was reported, although the relatively low melting temperature of about 2050°C was pointed out as a limitation. McQuillan reported successful use of aluminum oxide sheaths in the 1500-1900°C range.

Walker, Ewing, and Miller (1965, p. 816) conducted a thermocouple drift experiment where couples were sheathed with aluminum oxide insulators and fired at 1625°C in argon. After 72 hours at the firing temperature, drifts of –57, +42, and +38°C had occurred at the calibration temperatures of 860, 1200, and 1625°C, respectively. After additional exposure at 1625°C resulting in a total time of 120 hours there were only slight additional drifts. These investigators stated that the major initial contributor to drift was the tungsten; the molybdenum, which reacted with the aluminum oxide sheathing, produced a drift-contributing effect which was at first mostly offset by internal changes in the wire. After about 140 to 150 hours, however, the reaction involving the molybdenum became more rapid and the conductors reacted to destruction.

There was much development work on the W vs Mo thermocouple for a number of years, especially in England, for applications in measuring the temperature of liquid steel. Applications with higher melting point metals were also studied. Discussions of this type of use were given by Clark (1946), Sasagawa, et al. (1949), Potter and Grant (1949), Grant and Bloom (1950), Simons, et al. (1953), and Fedorovskii (1969).

W vs Mo-0.5Al. The addition of 0.5 per cent aluminum to molybdenum results in the marked change of output shown in Figure 5-2, compared with the W vs Mo thermocouple. The curve was first given by Adabaschjan (1959), who stated that the thermoelectric power was 11.5 μV/°C in the 1450-1700°C region. The output was low in the 0-100°C range so that reference junction compensation such as an ice bath was not required. Brief use up to 2000°C was permissible, so the couple was adaptable for liquid steel temperature measurements. Loss of aluminum results in a relatively rapid drift, and as a result the couple is unsuitable for long term measurements.

W vs Mo-1Fe. Schulze (1939) briefly discussed this pair and attributed the original development to W. Goedecke. The voltage-temperature curve in Figure 5-2 shows a polarity reversal near 400°C with tungsten positive at higher temperatures. Weber (1950) also mentioned this thermocouple and indicated that it can be used to 2200°C.

W vs Nb. This thermocouple was developed by Wilhelm and coworkers (1948) for use in vacuum and inert atmospheres during casting of high melting point metals. Calibrations were made up to 1772°C by the use of the melting point method and by comparison with a Pt-13Rh vs Pt thermocouple. Pletenetskii (1967) later investigated even higher temperatures. A voltage-temperature curve is shown in Figure 5-3, which was plotted from a table of average output values. A figure of 17.3 μV/$^\circ$C was given for the thermoelectric power at 1000°C.

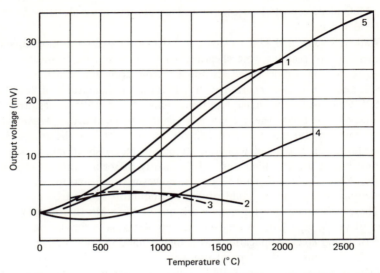

FIGURE 5-3   Tungsten paired with niobium and with tungsten-rhenium alloys.   1. W vs Nb, Pletenetskii (1967).   2. W-3Re vs W; 3. W-5Re vs W: Lachman and McGurty (1962, p. 177).   4. W vs W-10Re;   5. W vs W-20Re: Lachman and McGurty (1962, p. 153).

A batch to batch difference in properties of the niobium led to calibration differences of as much as 100°C, according to Wilhelm, et al. Within each batch or lot of wire, calibration differences were within ±25°C. Pletenetskii reported that for his work the niobium was annealed for 10 minutes in vacuum at 2200°C before calibrating. The anneal rendered the niobium soft and easily bendable. Heat treatment of the tungsten did not significantly change its thermoelectric properties so it was used without prior heat treatment. The calibration work was done in a vacuum and the repeatability of calibration points was 0.5 per cent or better.

A low thermoelectric power near room temperature eliminates the need for reference junction compensation when the cold junction temperature is below about 30°C. Aluminum oxide was recommended by Pletenetskii as the best insulation for this thermocouple. Stadnyk and Samsonov (1964) cautioned against the use of a helium atmosphere because of absorption of the gas by niobium with consequent voltage instability. Although Wilhelm, et al., used this couple in inert atmospheres, it is not known what limitations may exist for applications in gases other than helium.

W-3Re vs W. Lachman and McGurty (1962, pp. 153, 177) calibrated this pair while studying alloys of tungsten-rhenium for thermocouple application. The resulting voltage-temperature characteristics are included in Figure 5-3.

W-5Re vs W. The voltage-temperature curve, also from Lachman and McGurty, is given in Figure 5-3. Bennett, Hemphill, and Rainey (1966) found that this combination was stable in an atmosphere of helium with 100 ppm carbon monoxide up to about 1150°C. Above this temperature, increasing amounts of drift were observed over periods on the order of 50 to 200 hours.

W vs W-10Re. The calibration curve from the work of Lachman and McGurty (1962, p. 153) is shown in Figure 5-3.

W vs W-20Re. Lachman and McGurty also evaluated this combination. The output is relatively high at higher temperatures, as can be seen in Figure 5-3. The curve is nearly linear in the 1000-1500°C temperature range with an average thermoelectric power of about 17 $\mu V/°C$.

Novak and Asamoto (1966, 1968) described a stranded bare wire couple, with the cables enclosed in a W-26Re sheath and held under tension by molybdenum springs to prevent electrical shorting. Voltage fluctuations were observed during constant temperature conditions, caused either by high temperature creep or by vibration. A stranded wire thermocouple was also described for use in a special test furnace.

W vs W-(25-26)Re. Thermocouples of this type were developed during the late 1950, early 1960 period, when interest turned to the use of W-Re alloys for replacement of rhenium in W vs Re thermocouples. Davies (1960) published calibration data, followed shortly thereafter by Nadler and Kempter (1961), Lachman (1961), and Lachman and McGurty (1962, p. 177). Curves with some of these data are shown in Figure 5-4. Using the table of Lachman and McGurty, the thermoelectric power is 21.2, 18.2, and 17.2 $\mu V/°C$ at 1200, 1600, and 2000°C, respectively, for W vs W-26Re. Work by Williams, Veeraburus, and Philbrook (1968) gave a value of 19.73 ±0.05 $\mu V/°C$ over the 700-1100°C range for W vs W-25Re.

Some of the calibration work with this couple has been done at temperatures of 2760°C or higher. Small nonuniformities depending on the wire size of the tungsten have been reported. Also, initial instabilities during the first high temperature exposure of the wires can be expected. Lachman (1961) reported that an initial half-hour heat treatment of W vs W-26Re at 1600°C resulted in a 9°C increase of apparent temperature[3] at 800°C, but that subsequent thermal cycling to as high as 2300°C resulted in an additional change of only 2°C. Similar results were reported by Hall and Spooner (1963). Sata and Kiyoura (1963) found, however, that prior to heat treatment thermocouple output was unusually high above 2200°C, but decreased to a normal value after 2 hours at 2400°C. These investigators made calibrations accurate to ±3°C by the melting point method up to the melting point of iridium at 2447°C, and also calibrated by optical pyrometry.

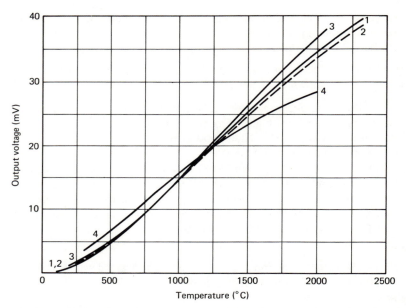

FIGURE 5-4   Tungsten vs rhenium and tungsten vs tungsten-rhenium alloy thermocouples.   1. W vs W-25Re, Nadler and Kempter, (1961).   2. W vs W-26Re, Lachman and McGurty (1962, p. 177). 3. W vs W-30Re, Sims, Gaines, and Jaffee (1959).   4. W vs Re, Thomas (1963).

These thermocouples can be used in inert atmospheres, in hydrogen, or in a vacuum. Lachman (1961) described several drift tests in an argon atmosphere including a 500-hour exposure at 982°C which resulted in a positive drift of about 2°C. Bennett and Hemphill (1963), and Bennett, Hemphill, and Rainey (1966) stated that total drift was within ±10°C after 950 hours at 1750°C in a helium atmosphere. Hendricks and McElroy (1966) found similar results at somewhat lower temperatures with sheathed couples in vacuum. Work by Burns and Hurst (1971) indicated that bare wire thermocouples of this type would show no significant drift during a 500-hour exposure at 2127°C in argon. In view of their results with 1000-hour exposures of W-3Re vs W-25Re it is likely that the 500-hour drift-free period can be considerably extended. Further, similar performance can be expected in atmospheres of helium, hydrogen, and nitrogen. Capability in vacuum is likely to be more limited because of preferential evaporation of rhenium and accompanying drift.

The presence of impurities in an otherwise inert atmosphere may result in drift. Oxygen present in the amount of 100 ppm was observed by Bennett and Hemphill (1963) to result in instability, followed by complete failure after a few hours. When the oxygen content was only 10 ppm there was a significant positive drift at an exposure temperature of 1200°C, resulting in a 34°C deviation after 2 hours. Carbon monoxide in amounts of 25 ppm and 100 ppm also produced large amounts of drift. The amount of drift was dependent on wire size and temperature, with a rapid drift for the first 100 hours followed by a subsequent slower drift. Drift rates were similar for both impurity amounts.

Nadler and Kempter (1961) found that W vs W-25Re couples increased in output in the presence of graphite, and in a helium atmosphere. The output increased by 2 to 3 per cent after 15 minutes at 2300°C. Hall and Spooner (1964, 1965) described tests with W vs W-26Re, in argon with carbon vapor, and indicated that drift would not exceed ±2 per cent during 100 hours at temperatures up to 2400°C. Drift was greater at decreased atmospheric pressures because of increased carbon vapor transport and subsequent reaction with the thermocouple. When a hydrogen atmosphere was used with carbon vapor present, drift occurred within a few hours at 1800°C. Drift was even more pronounced when methane was used in place of hydrogen. Farrow and Levitt (1964) also reported thermocouple drift during tests with carbon vapor present at temperatures up to 2400°C and with argon pressures to as low as 0.2 torr. Williams, et al. (1968) found that W vs W-25Re was stable to at least 1100°C in the presence of graphite, and also found that the lean sulphur atmospheres produced by sulfide melts did not affect stability. Tellurium vapor attacked and completely destroyed exposed tungsten wires, however.

McGurty and Kuhlman (1962) made detailed studies of the electrical resistance and thermoelectric effect of the following high purity materials in contact with W vs W-26Re conductors; Al_2O_3, BeO, CaO, HfO_2, ThO_2, Y_2O_3, and ZrO_2. Beryllium oxide was found to be the most satisfactory of these. Meservey (1967) showed data indicating that boron nitride has a high temperature resistivity that is intermediate between that of thorium dioxide and beryllium oxide, but he observed thermocouple instability and drift above 2000°C in the case of boron nitride.

Walker, Ewing, and Miller (1965, p. 816) investigated the drift of aluminum oxide-sheathed W vs W-26Re thermocouples at 1625°C in argon. A 120-hour exposure resulted in a 41°C increase of apparent temperature. Much of this increase occurred during the first 20 hours when a drift of 25°C was observed. Heat treatment prior to testing would have decreased the initial drift, but at the expense of wire embrittlement. The tungsten alloy wire was less stable than the tungsten.

Funston and Kuhlman (1967) investigated the behavior of various varieties of insulated W vs W-25Re thermocouples at high temperatures. Use of hafnium dioxide resulted in an unusually high output. Kuhlman and Baxter (1969, p. 319; 1969, Oct.) conducted drift tests with thermocouples enclosed in hafnium dioxide two-hole insulators produced from material of spectrographic grade purity; these assemblies were sheathed by W-25Re tubing. During a 1000-hour exposure in helium at 2300°C a downward drift occurred resulting in a -200°C change of apparent temperature. It was believed that this behavior was caused by traces of oxygen in the helium atmosphere rather than incompatibility of the materials that were used.

Improved sheathed thermocouples with beryllium oxide insulators were developed by Goodier, Fries, and Tallman (1971). The oxide purity was at least 99.8 per cent; a special insulator shape and method of wire placement were used to lengthen possible shunting paths between the wires and to accommodate thermal expansion effects. Particular attention was also paid to cleanliness during fabrication. It was reported that no shunting errors due to decreased insulator resistivity were observed up to 2027°C. These couples were satisfactory for thermal cycling to 1777°C. An accuracy of 1 per cent was maintained during 50 thermal cycles between 177 and 1777°C. Usage in an environment of hydrogen with the presence of carbon led to the choice of molybdenum for the sheath, since it does not react with hydrogen. Also, a tantalum inner liner was used to retard carbon diffusion to the insulators and the wires.

Coaxial thermocouples, and metal sheathed thermocouple probes and cables have been described by several investigators, including Lachman (1962), Gross, et al. (1964), Clark (1964), Hoppe and Levine (1966), Briggs and Johnston (1966), and Gross, et al. (1966). Aluminum oxide, beryllium oxide, hafnium dioxide, and thorium dioxide have been used as insulation with various types of refractory metal sheaths, as well as sheath coatings for additional protection.

Nuclear applications were reported by Driesner and coworkers (1962; 1962, May), Prince and Sibbett (1965), Kjaerheim (1965), Christenson and Ferrari (1966), and Wood (1967). Heckelman and Kozar (1971) reviewed work on these and similar thermocouples in their paper on irradiation studies of W-3Re vs W-25Re. Funston and Kuhlman (1966) gave quantitative drift results for a nuclear environment which showed that irradiated couples drifted in the positive direction at low temperatures and in the negative direction at higher temperatures. Transmutation effects resulted in a +13°C error at 600°C and a -15°C error at 1900°C after exposure to an integrated thermal flux of 6.2×10^{20} n/cm^2, and an integrated fast flux of 1.1×10^{20} n/cm^2.

Kuhlman (1966) also studied transmutation influence on thermocouple output by synthesizing homogeneous alloys with the computed amounts of contaminants which would exist after given amounts of irradiation. Osmium is the principal foreign element which appears during transmutation. Skaggs, Ranken, and Patrick (1967) found, however, that during actual irradiation the transmutation was accompanied by phase separation, so that two alloy phases were present in the W-25Re conductor, one containing a much higher proportion of osmium than the other. The resulting thermocouple drift was less than predicted drift based on the voltage-temperature characteristics of synthesized alloys, presumably because no phase separation was present in the latter.

Skaggs, et al., found from their irradiation tests of W vs W-25Re that an increase of 60°C in apparent temperature at 927°C and 33°C at 1897°C occurred, during an exposure of 5328 hours at an average temperature of 1727°C; the thermal flux was 2.0×10^{13} n/cm^2 sec and the total exposure was 3.8×10^{20} n/cm^2. The thermocouple drift was caused by the combined effects of transmutation and phase separation. Both transmutation and temperature-time history influenced the amount of calibration change.

W vs W-30Re. Sims, Gaines and Jaffee (1959) developed a voltage-temperature characteristic for this combination based upon calibrations of W vs Re and Re vs W-30Re. The curve is shown in Figure 5-4 and is nearly linear over a wide range of temperatures. The thermoelectric power is about 22 μV/°C between 1000 and 2000°C. There has been little additional work done on this or other thermocouples where tungsten is alloyed with rhenium as a solute in amounts exceeding 26 per cent. According to Lachman and McGurty (1962, pp. 153, 177) there are fabrication difficulties associated with tungsten alloy wires of greater than 26 per cent rhenium. Also, the solubility limit lies very near the 30 per cent figure. Kuhlman (1963, Sept.) reported that a two-phase mixture resulted during an attempt to prepare this alloy.

W vs Re. The first results with W vs Re thermocouples have been generally credited to Goedecke (1931); the work was limited in scope, however. Haase and Schneider (1956) later used tungsten as a reference metal in an investigation of the iridium-rhenium alloy system and also obtained the voltage-temperature characteristic of W vs Re. Sims, Craighead, and others (1956) described a calibration at about the same time, following general studies of rhenium by Sims and his associates. Sims, Gaines, and Jaffee (1959) calibrated this couple from 0 to approximately 2200°C. Comparisons of the results of Haase and Schneider, unpublished work by G. Rouse of the National Bureau of Standards, early work by J. Lachman, and their own results, were included. Nadler and Kempter (1961), and Lachman and McGurty (1962, p. 177) also presented voltage-temperature data, and Caldwell (1962) included additional information in his review article. Thomas (1963) at the National Bureau of Standards shortly afterward published the present reference tables for this thermocouple. A curve from the latter reference is shown in Figure 5-4. According to Thomas, the thermoelectric power S is 17.6, 17.6, and 13.1 μV/$^{\circ}$C at 500, 1000, and 1500°C, respectively. At higher temperatures S declines and is only 6.0 μV/$^{\circ}$C at 2000°C.

Thomas calibrated 11 thermocouples, made from 8 different lots of tungsten wire from 3 manufacturers. The rhenium wire was obtained from the sole producer of that material in the United States. The tables cover the range of 0 to 2000°C. The greatest spread in voltage that occurred among the 11 thermocouples in the 0 to 1000°C range was the equivalent of 15.2°C, or 4.1 per cent of the calibration temperature, which was 370°C. Between 1000 and 2000°C the corresponding difference was 44°C at 2000°C or 2.2 per cent. The large percentage variation or spread at 370°C was considered to be rather surprising, and was due primarily to variations in the tungsten wires. Instabilities were observed, and heat treatment was used at a higher temperature than a given data point temperature, prior to making a measurement, in order to minimize them. Differences in voltage between tungsten wires of various diameters were also found.

The small instabilities in tungsten, rhenium, and W vs Re thermocouples, which are caused by initial heating and by additional short term heat treatment, were described by Kuether (1960; 1962, pp. 229, 233, 149), Lachman and Kuether (1960, pp. 31, 67), and Kuether and Lachman (1960). As with other thermocouples utilizing tungsten, a high temperature heat treatment should be used prior to calibration whenever possible to avoid later changes in the voltage-temperature characteristic. These changes are associated with stress relief and recrystallization that occur during the first heating cycle. Where possible the thermocouple should be installed before heat treatment to avoid handling the embrittled tungsten.

One of the tests described by Lachman and Kuether consisted of heating an untreated thermocouple for 5 minutes at 2000°C and observing the change in the 1000°C calibration point, which amounted to +4°C.[4] An additional 4 hours at 1500°C resulted in a further 3°C increase of apparent temperature at 1000°C, but after that measurement there was no further drift caused by stress relief or recrystallization, since such processes were virtually complete after the lengthy heating period.

Kuether (1962, p. 149) discussed inhomogeneity of the W vs Re thermocouple, which is caused almost entirely by the rhenium. Inhomogeneities in a typical thermocouple resulted in ±3.7°C errors with a hot junction temperature of 982°C, ±9.4°C errors at 1482°C, and led to ±44°C errors at 1982°C. Above 2200°C the inhomogeneity errors of W vs Re couples were so large that temperature measurements in that region could not be recommended. The inhomogeneities producing these results were in sections of wire of some 20 cm in length. A second factor limiting high temperature applications was pointed out by

Lachman and McGurty (1962, p. 153) to be the significant decline of thermo-electric power in the 2200°C region. As a result of these two limitations the maximum recommended application temperature is usually given as 2000 to 2200°C, even though the metals can withstand higher temperatures.

In general, W vs Re can be used in vacuum, in inert atmospheres, or in hydrogen. Kuether and Lachman (1960) found no significant drift during 20 hours of vacuum exposure at temperatures up to 1427°C. Much of the calibration work by the various investigators was done in a helium atmosphere. Drift data are available for exposure to hydrogen from the results of Lachman and Kuether (1960, pp. 31, 67). There a previously heat treated couple showed no significant drift, as measured at 1000°C, after 9 hours at temperatures ranging from 1800 to 2100°C.

Results when contaminants were present in otherwise inert atmospheres were described by Nadler and Kempter (1961), Bennett and Hemphill (1963), and Bennett, Hemphill, and Rainey (1966). As with W vs W-26Re thermocouples, W vs Re couples showed large drifts when carbon monoxide in amounts of 25 or 100 ppm was present. Nadler and Kempter (1961) concluded that short time operation in helium with the presence of graphite at temperatures above 2100°C was feasible.

The performance of as-received (not heat treated) thermocouple wire in aluminum oxide insulators was described by Walker, Ewing, and Miller (1965, p. 816) for a temperature of 1625°C in an argon atmosphere. An initial instability was noted during the first hour, which was due to the tungsten wire. This corresponds to the initial stress relief and recrystallization process. There was little additional change for about 150 hours; that which did occur was also attributed to recrystallization and other internal changes in the thermocouple. Drift after 20 hours was +20°C and increased to a net total of +27°C apparent temperature change after 120 hours. After about 150 hours the dominant process was metal reaction with the aluminum oxide which then proceeded rapidly.

Tungsten versus rhenium thermocouple probes were described for measurement of liquid steel temperatures by Ramachandran (1963), Gross and Griffith (1963), and Ramachandran and Acre (1964). A nuclear application was described by Trauger (1960), and similar work was discussed by Kjaerheim (1965) and by McQuilkin, et al. (1965).

W vs Ta. Early calibration work was described by Morugina (1926). Roeser and Wensel (1941, p. 1293) later calibrated the individual metals referenced to platinum, from which data on W vs Ta may be computed. Other calibration results have been reported by Wilhelm, et al. (1948), Lander (1948), Morgan and Danforth (1950), Troy and Steven (1950), Lachman and McGurty (1962, p. 177), and Feigenbutz (1968). The calibration curve of Morgan and Danforth is shown in Figure 5-5 for high temperatures, and the data from Roeser and Wensel are plotted for the lower temperature region. The average thermoelectric power

S in the 1200-1300°C interval is 13 $\mu V/°C$. At higher temperatures S decreases, and the thermocouple output goes through a maximum in the 2000°C region. Although Morgan and Danforth extended their calibration to 3000°C, it was reproducible only to 2000°C, due to causes which were not determined.

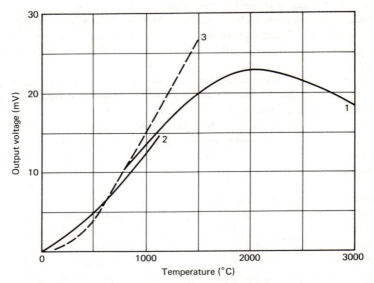

FIGURE 5-5   Tantalum and titanium paired with tungsten.   1. W vs Ta, Morgan and Danforth (1950).   2. W vs Ta, Roeser and Wensel, (1941, p. 1293).   3. W vs Ti, Stadnyk and Samsonov (1964).

There are significant differences between the various voltage-temperature curves which have been reported. The degree of agreement shown between the two curves in Figure 5-5 is better than usual for the tungsten-tantalum thermocouple. The lack of agreement is generally considered to be a result of the great sensitivity of tantalum to internal impurities and to contamination by additional impurities in the atmosphere that is used. Although these difficulties were apparently overcome for a few specific applications, other investigators have concluded that this thermocouple is unsatisfactory. It is definitely not a recommended choice for the inexperienced user.

Lander (1948) reported successful application in a hydrogen atmosphere at 5 torr; Troy and Steven (1950) suggested that the wires were actually protected from the hydrogen, however. The latter investigators reported a fair amount of stability at temperatures up to 1300°C in a vacuum, and they described the gradual contamination of tantalum in a vacuum system controlled by a mechanical pump, because of impurities in the residual atmosphere. Negative drift of a W vs Ta thermocouple was noted during operation at 1500 and 1600°C.

Morgan and Danforth's calibration work was done in a vacuum of 10^{-5} torr and wires from different spools and of different sizes were tested, apparently without observing large deviations or nonuniformities. Walker, Ewing, and Miller (1965, p. 816) observed considerable instability of tantalum at relatively low temperatures and did not recommend its use in thermocouples. Reaction with aluminum oxide in an atmosphere of argon at 1625°C was also described which resulted in rapid failure of the wire. Wires electrically heated in argon also failed after a relatively short period of time and were more brittle than wires which were heated in vacuum. The latter were heated at 1625°C and did not fail during 24 hours of exposure.

Feigenbutz (1968) used W vs Ta in an unusually low temperature range for this combination, making a calibration in the 0 to 420°C interval. Tungsten and tantalum were selected because of the very low solubilities of these metals in mercury; the thermocouple was used for direct temperature measurements of boiling mercury.

W vs Ti.   Stadnyk and Samsonov (1964) included a voltage-temperature curve of this combination in a review article. The curve is shown in Figure 5-5. No details or references were given.

Tungsten Alloy Thermocouples

W-3Re vs W-15Re.   This combination was developed by Danishevskii (1969) in the 1950's for furnace temperature measurement applications, and was chosen because the output up to 50-60°C was similar to that of Pt-10Rh vs Pt. This permitted existing extension wires for the noble metal couple to be used with the high temperature thermocouple. The designation for this pair in the Soviet Union is type VR 3/15. The voltage-temperature curve[5] is shown in Figure 5-6. Danishevskii also gave a curve of thermoelectric power versus temperature, showing values of 12.5, 12.5, and 10.2 μV/°C at 600, 1000, and 1600°C, respectively.

W-3Re vs W-20Re.   A calibration curve from Lachman and McGurty (1962, pp. 153, 177) is given in Figure 5-6. From the curve, the thermoelectric power is 12.6 μV/°C in the 1650°C region. This was included with other thermocouples in work by Lachman and McGurty on the tungsten-rhenium alloys. The general shape of the voltage-temperature curve is quite similar to that of W-5Re vs W-26Re, but the output is lower.

W-5Re vs W-20Re.   Danishevskii (1969)[6] calibrated this thermocouple and considered it preferable to W-3Re vs W-15Re because the 5 per cent rhenium alloy retained greater ductility after heat treatment than the alloy with 3 per cent rhenium. Danishevskii, Ipatova, et al. (1964) showed a calibration curve between –200 and +2500°C, and gave values of thermoelectric power of 8.0 and 14.0 μV/°C at –196 and +100°C, respectively. At higher temperatures, figures of 16.0, 13.0, and 11.0 μV/°C were given, corresponding to temperatures of 1000, 1500, and 2000°C. The voltage-temperature curve in Figure 5-6 was taken from a table of average values prepared by Danishevskii, Oleinikova, et al. (1968). Uniformity of available wires in the Soviet Union was such that thermocouples generally had characteristics within ±1 per cent of the values in the table. The Russian designation for this thermocouple is type VR 5/20.

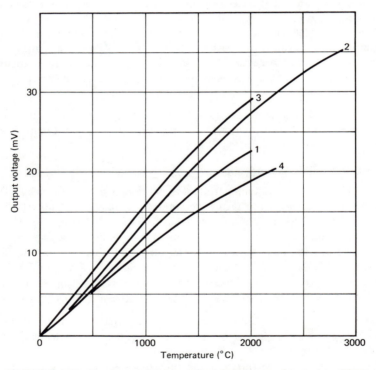

FIGURE 5-6   Tungsten-rhenium alloy thermocouples, I.   1. W-3Re vs W-15Re, Danishevskii (1969).   2. W-3Re vs W-20Re, Lachman and McGurty (1962, pp. 153, 177).   3. W-5Re vs W-20Re, Danishevskii, Oleinikova, et al. (1968).   4. W-10Re vs W-20Re, Danishevskii, Ipatova, et al. (1964).

According to Danishevskii, Ipatova, et al., drift in vacuum or in hydrogen was within ±25°C over periods of up to several hundred hours at temperatures of 1400 to 1800°C. A change of 30-35°C was found after operating for 400 hours at 1400°C in argon. It was stated that in general this combination can be used in atmospheres of argon, nitrogen, helium, and hydrogen, as well as in a vacuum or in the presence of carbon.

Pletenetskii and Mandrich (1965) described drift in helium at 1500°C for a couple which had previously been heat-treated for 3 minutes at 1800°C to improve the initial stability. During heating at 1500°C the output first drifted negative by 2.1°C after some 20 hours and then began drifting in the positive direction. After 250 hours the apparent temperature showed a 7°C positive drift.

Danishevskii, Ipatova, et al. (1964) indicated that measurements were possible while in contact with titanium and molybdenum. Lapp and Popova (1964) found that a one-hour exposure at 1500°C in the presence of molybdenum had little effect, but that drift resulted at 2100°C. This was caused by sensitivity of the W-20Re alloy when in direct contact with the molybdenum, or when within but not touching a cylinder of that metal.

Lakh and coworkers (1963) concluded that operation in vacuum, inert atmospheres, or hydrogen was satisfactory for bare wires at temperatures up to 1400°C, but recommended aluminum oxide insulation as protection at higher temperatures. In carbon-contaminated atmospheres up to 2000°C a protective jacket of beryllium oxide was suggested. Finally, preference was expressed for alloys of 10 or 20 per cent rhenium rather than W-5Re for use at the higher temperatures. Later high temperature work by Oleinikova, et al. (1969) in an atmosphere of argon confirmed that W-10Re vs W-20Re performed better than W-5Re vs W-20Re at high temperatures. It was suggested that performance of the latter couple could be improved by additional preliminary heat treatment of the wires.

Terebukh, et al. (1968) described a special application where a thermocouple probe was used in temperature measurements during the fire refining of copper at 1150°C. Ergardt (1968) stated that errors with probes for liquid steel temperature measurement using W-5Re vs W-20Re and similar tungsten-rhenium alloy thermocouples did not exceed the error associated with platinum-rhodium versus platinum thermocouples. Vedernikov (1969) used this couple in determining the thermoelectric powers of transition metals at high temperatures. The work was done in vacuum and in a helium atmosphere.

W-10Re vs W-20Re. Danishevskii, Ipatova, et al. (1964) presented a voltage-temperature curve which is shown in Figure 5-6. A table of thermoelectric power data was also given, showing values of 6.0 and 10.0 μV/°C at –196 and +100°C, respectively. At higher temperatures values of 10.0, 8.0, and 7.0 μV/°C were given for 1000, 1500, and 2000°C. As shown in Figure 5-6 the output is lower

than that of W-5Re vs W-20Re. The W-10Re vs W-20Re thermocouple is designated as the VR 10/20 couple in the Russian literature.

This thermocouple can be used in the same atmospheres and for the same
general applications as W-5Re vs W-20Re, but gives better performance at 2000°C
and higher. Oleinikova, et al. (1969) exposed as-received wires to a temperature
of 1800°C, for 20 hours, in an argon atmosphere. The resulting drift was a decrease of 1 per cent in output. At 2000°C a similar test resulted in a drift of 2 per
cent, the direction not being specified. A W-5Re vs W-20Re couple drifted 6 per
cent under like conditions. The greater part of the drift for both couples occurs
during the first few hours, because of initial instabilities in the untreated wires.

W-15Re vs W-20Re.   Lakh and coworkers (1963) recommended the higher
rhenium content tungsten alloys for use in hydrogen, this combination being
particularly suitable. Positive drift was about 10°C during exposures of up to
275 hours at 1400°C and 230 hours at 1800°C. Calibration data were shown for
the 1370-1440°C range under these conditions. Stadnyk and Samsonov (1964)
also recommended this couple for hydrogen.

W-3Re vs W-25Re.[7]   Zysk, Toenshoff, and Penton (1963) evaluated couples
between 0 and 2400°C and prepared a tentative calibration table. The curve in
Figure 5-7 was constructed using the tabulated data. Calibration procedures and
a summary of the revised version for the table were given by Zysk and Toenshoff
(1967), and were later converted to the International Practical Temperature Scale
of 1968 by Tseng, Schnatz, and Zysk (1970). The complete table of millivolt
outputs in one-degree increments from 0 to 2400°C is given in the latter reference. The thermoelectric power is 19, 17, and 14 $\mu V/°C$ at 1200, 1600, and
2000°C, respectively.

Fanciullo (1964) also did calibration work, and the calibration range was
extended to 3000°C by Asamoto and Novak (1967). The latter investigators
found that data from individual thermocouples were within ±1 per cent of the
best-fit curve for the 200-2000°C range, and were within ±2 per cent for the
2000-3000°C interval. The thermoelectric power is only about 6 $\mu V/°C$ at
3000°C, but that is still an adequate value for practical applications. The curve
from Asamoto and Novak is included in Figure 5-7.

The W-3Re vs W-25Re thermocouple is similar to the W-5Re vs W-26Re
combination in terms of voltage-temperature characteristics. The manufacturer[8]
of the former combination claims better ductility after heat treatment by use of
specially prepared W-3Re wire. Normally the alloys with higher amounts of
rhenium would be more ductile. In this case, however, the W-3Re is the same
type of wire that is used in incandescent lamp filaments and which was originally
developed for improved sag resistance. The metal powder for the wire is said to
be doped with small amounts of aluminum, potassium, and silicon compounds
to influence the microstructure of the wire which develops during its manufacture

by powder metallurgical methods. During the fabrication process the volatile additives are almost entirely eliminated and the altered microstructure results in improved ductility. Zysk and Robertson (1971) reviewed this topic and included a number of references on the subject of doping agents and their effects upon

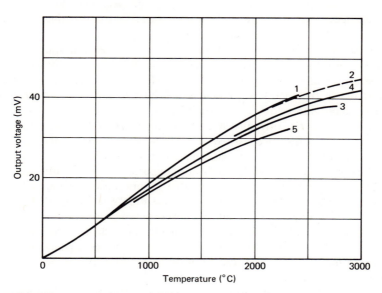

FIGURE 5-7 Tungsten-rhenium alloy thermocouples, II. 1. W-3Re vs W-25Re, Zysk, Toenshoff, and Penton (1963). 2. W-3Re vs W-25Re, Asamoto and Novak (1967). 3. W-5Re vs W-26Re, Lachman and McGurty (1962, p. 177). 4. W-5Re vs W-26Re, Asamoto and Novak (1967). 5. W-7Re vs W-26Re, Lachman (1961).

Novak and Asamoto (1965, 1966, 1967) exposed bare wire thermocouples to a temperature of 2600°C following calibration to 2900°C in vacuum, and reported that during a period of 15 hours drift was within ±2 per cent. An investigation of initial instability after calibration, and of drift over periods up to 1000 hours, was reported by Burns and Hurst (1970, 1971). Bare wire thermocouples were used, and the entire lengths of the wires were exposed at the same temperature except when calibrations were made. Tests were conducted in atmospheres of argon and hydrogen, and also in a high vacuum of less than 10^{-8} torr. In all three cases an initial period of calibration change was observed during high temperature exposure of wires which had not been previously heat treated. The change was complete in less than one hour at 2127°C and was probably complete in much less than 50 hours at 1927°C. The apparent change in temperature at 1800°C was positive and less than 20°C in all cases.

Similar results have been reported for other types of tungsten-rhenium thermocouples. The condition appears to be caused, at least in part, by stress relief and recrystallization. Burns and Hurst found, however, that under some conditions additional recrystallization and grain growth occurred long after the initial changes were complete, but without significantly affecting the thermocouples.

Zysk and Robertson (1971) stated that an improved annealing process can be used to produce wires that show initial instabilities on the order of 0.1 to 0.2 per cent while still retaining ductility. No details were given. Burns and Hurst stated in their 1971 paper that proper heat treatment of as-drawn wire[9] can essentially remove the initial instability of the characteristics and yet leave the conductors sufficiently ductile at room temperature for sensor fabrication purposes. Their treatment consisted of a one-hour exposure in argon at 2127°C for W-3Re and a few minutes at that temperature for W-25Re.

Burns and Hurst found that additional exposures at temperatures of as high as 2127°C in argon did not result in further calibration changes after the initial change had occurred. No significant drift was observed during 1000 hours at 2127°C. Similar results were found for wires exposed over periods of up to 1000 hours in nitrogen and for 500 hours in helium or hydrogen. In high vacuum there was significant drift, which was caused by preferential evaporation of rhenium from the alloys. The amount of drift was critically dependent on temperature. Drift was negligible after 500 hours at 1927°C; a negative drift of more than –200°C, measured at 1800°C, was reported after only 50 hours exposure at 2327°C.

Walker, Ewing, and Miller (1965, p. 816) evaluated W-3Re vs W-26Re couples,[10] as well as a number of other types, during a study of thermocouple drift. It was found that the W-26Re alloy was the major contributor to drift when paired with tungsten, rhenium, or the dilute rhenium alloys of tungsten. In a test at 1625°C in argon, with the presence of aluminum oxide sheathing, a 72-hour exposure time resulted in an 18°C apparent temperature increase. After 120 hours the drift had reached 24°C.

Briggs and Johnston (1966) tested thermocouples enclosed in beryllium oxide insulators for a period of 610 hours, at about 1920°C, in an atmosphere of argon. The total drift was less than 1 per cent of the test temperature. Fitts, Miller, and Long (1971) operated similar thermocouples for up to 125 hours at 2000°C and to 500 hours above 1800°C with calibration changes of less than 5 per cent. The wires and the beryllium oxide sheaths were observed to have developed a degree of porosity under these conditions, and vapor transport effects involving the metals and the insulation were also described.

Novak and Asamoto (1965, 1966, 1967), and Asamoto and Novak (1967) studied the high temperature behavior of sheathed thermocouples. Beryllium oxide and thorium oxide insulations were used. The former did not affect thermocouple output at 2300°C, but with thorium dioxide there was a 3 per cent decrease of thermocouple output compared with bare wire calibrated output. This was apparently caused by the shunting effect of the oxide as it became partially conductive at high temperature. The shunting effect was reproducible within ±1 per cent. At 2700°C the apparent temperature decrease was 7 per cent and was also reproducible. Vitrified thoria insulators were used in at least some cases, and in two instances were sheathed with W-26Re tubing. A review of previous work from other sources was given, where use of powdered thorium dioxide insulation had resulted in an extreme shunting effect and in large deviations from the bare wire calibration curve. An insulator purity range from 99.0 to 99.9 per cent made little difference in thermocouple performance, in the case of Novak and Asamoto.

Kuhlman and Baxter (1969, Oct.) conducted an 1100-hour drift test with couples in high purity thorium dioxide two-hole insulators and W-25Re sheaths. The temperature was 2300°C and a helium atmosphere was used with an oxygen content in the 1 to 2 ppm range. Drift was 50°C or less in magnitude and in the negative direction during the period of the test.

Burns and Hurst (1971) tested three thermocouples insulated with beryllium oxide tubing of at least 99.8 per cent purity, while in the presence of tantalum. A temperature of 1800°C was maintained for 1029 hours in an argon atmosphere, and positive drifts of only 2.4, 2.6, and 3.0°C were observed for the three couples. Work by Droege, et al. (1971) suggests that the high purity of the insulators was an important factor in obtaining this result, because of drift which might otherwise occur with tantalum present. Burns and Hurst gave numerous details concerning the composition and preparation of the insulators for use.

Heckelman and Kozar (1971) reported on tests of swaged thermocouples with beryllium oxide insulation, tantalum sheaths, and grounded hot junctions. The purity of the beryllium oxide exceeded 99.93 per cent; an atmosphere of helium was used. A 5000-hour exposure at about 1700°C resulted in a drift of –54°C. A temperature-time history of the thermocouples showed additional times at intervals during the test when some 16 thermal cycles between about 370°C to the high temperature region were also conducted.

Popper and Knox (1967) included discussion of this thermocouple in a review of work oriented for nuclear applications. Novak and Asamoto (1968) described a number of experiments with bare wire and sheathed thermocouples in various reactor tests.

Heckelman and Kozar (1971) conducted irradiation tests on couples of the same type as were used for the drift tests described earlier. These thermocouples were exposed for a 5000-hour period at about 1700°C in a helium-argon environment. The integrated thermal flux was 2×10^{21} n/cm^2 and the integrated fast flux was 0.4×10^{21} n/cm^2 for energies greater than 0.18 meV. The negative drift component due to irradiation effects was $-136°C$. The total drift including non-nuclear effects was $-190°C$.

The same type of thermocouple was tested by Carpenter, et al. (1971) for 4470 hours at about 975°C in a helium atmosphere, except when brief in situ calibrations were made by the melting point method using copper at 1083°C. The integrated thermal flux during the test reached 1.6×10^{21} n/cm^2. Similarly, the integrated fast flux for energies exceeding 0.18 meV reached 2.7×10^{21} n/cm^2. Total drift due to all environmental effects was the equivalent of $-85°C$, measured at 1083°C.

W-5Re vs W-26Re.   The experience of a number of investigators with the tungsten-rhenium thermocouple indicated that good performance at high temperatures was possible. However, the much decreased thermoelectric power and the inhomogeneity of W vs Re at such temperatures encouraged study of the tungsten-rhenium alloys in a search for a couple useful at temperatures above 2000°C. The W-5Re vs W-26Re couple was found to be one of the best combinations for a satisfactory thermoelectric power, combined with improved homogeneity, uniformity, and ductility. The first extensive work with these alloys was reported at the 1960 Electrochemical Society Symposium on Rhenium by Lachman and McGurty (1962, p. 153). Additional reports were published shortly afterwards: Lachman (1961) and Lachman and McGurty (1962, p. 177). A calibration curve from the latter reference appears in Figure 5-7 for the 0-2800°C range. Tentative calibration tables by Lachman (1961) covered the 0 to 2320°C range, from which the thermoelectric power was found to be 18, 16, and 10 $\mu V/°C$ at 1000, 1500, and 2300°C, respectively.[11]

Later calibration work was done by Fanciullo (1964); Asamoto and Novak (1967) extended the upper end of the calibration range to 3000°C. A curve plotted from the latter work is included in Figure 5-7. The thermoelectric power is about 6.7 $\mu V/°C$ at 2800°C, a low but adequate value for applications work. The experimental data compared closely with averaged values for the 200-3000°C range, matching a best-fit equation to ±1 per cent from 200 to 2000°C and ±2 per cent from 2000 to 3000°C.

Heat treatment prior to usage was discussed by Lachman (1961), and Hall and Spooner (1963). These researchers pointed out that by replacing the tungsten wire in W vs W-26Re with W-5Re, both wires can then be heat treated to some degree before installation without complete loss of ductility. A brief high temperature heat treatment does not embrittle the wires, although both time and temperature are important parameters in this respect. Lachman reported that a 15-minute exposure at a temperature as high as 1700°C did not cause embrittlement of W-5Re, but such did occur after heating at a little above 2300°C.

It was also found that wires that were delivered in the ductile condition showed greater instability than that of W vs W-26Re, during the initial heating cycle. In some instances, additional heat treatment did not reduce instability below that of W vs W-26Re. However, Hall and Spooner also described evidence where heat treatment at just below the recrystallization temperature of W-5Re does reduce instability more effectively in the alloy than in tungsten, the recrystallization temperature of which is much lower. Tungsten would require a long exposure at the lower temperature to obtain heat treatment induced stability equivalent to that of W-5Re at the higher temperature.

It appears from the above information that a W-5Re vs W-26Re thermocouple can be installed in a ductile or semiductile condition after some degree of preliminary heat treatment and be calibrated without additional heat treatment if necessary. Nevertheless, for best performance, heat treatment of the installed wires to a temperature above the maximum use temperature is still good practice, although the wires may become quite brittle during the process.

This type of thermocouple can, in general, be used in a vacuum, inert atmospheres, or hydrogen. Hall and Spooner indicated that loss of rhenium from both conductors caused by evaporation at high temperature produces thermoelectric power changes in each alloy that tend in part to offset one another. This can be important in high temperature applications.

Thermocouple drift in argon of about –1°C per 1000 hours was reported by Hall and Spooner for 1000 to 4000 hour periods at temperatures of about 1100 to 1370°C. This drift was caused by a change in the W-26Re wire. There was essentially no change in the positive alloy. Exposure in hydrogen at 1540-1760°C was also described, with drift of less than 1 per cent during periods up to 8000 hours.

Glawe (1970), and Glawe and Szaniszlo (1971) found comparable drift for long term tests of thermocouples in argon. The couples were in aluminum oxide insulators of purity that exceeded 99.5 per cent and were held for 10,000 hours at 1327°C. Total drift after this time ranged from –21.1 to –22.5°C. From this the average drift was about –22°C per 10,000 hours. Similar tests were conducted in vacuum but were terminated after 3500 to 3900 hours. The thermocouples performed in the same manner as in argon up to that time. This led to estimates that a 10,000-hour operating period with drift of –22°C can be expected of 0.51 mm diameter thermocouple wires in either environment.

Work by Burns and Hurst (1970, 1971) indicated that thermocouples of this type would show no significant drift during a 500-hour exposure at 2127°C in argon. In view of their results with 1000-hour exposures of W-3Re vs W-25Re it is likely that the 500-hour drift-free period can be considerably extended. Further, similar performance can be expected in atmospheres of helium, hydrogen, and nitrogen.

All of these results indicate that good long-term performance can be expected of the W-5Re vs W-26Re thermocouple. Glawe and Szaniszlo (1971) pointed out from their results that this thermocouple has the same average

10,000-hour drift at 1327°C in argon as do noble metal couples of Pt-13Rh vs Pt.[12] In fact, at 10,000 hours individual deviations from the average drift were considerably less for the refractory metal couples than for the noble metal units. The expected average 10,000-hour drift is about the same for the two types of thermocouples in vacuum, but 0.51 mm diameter wires of the W-5Re vs W-26Re type have a full 10,000-hour capability compared with 2000 to 4000 hours for Pt-13Rh vs Pt of the same size. On the other hand, the initial drift of the noble metal type is low, so it is superior for total use times of 1000 to 2000 hours. It should be remembered that these figures apply to couples that are enclosed in high purity aluminum oxide insulators.

Novak and Asamoto (1965) found that bare wire thermocouple drift was less than ±2 per cent during 15 hours at 2600°C in vacuum. Carroll and Reagan (1966) detected no drift during exposure to an inert gas containing 3 per cent hydrogen at temperatures up to 1450°C. However, this thermocouple was shown to be sensitive to the presence of oxygen in amounts as small as 1 ppm over a period of several weeks at 1000°C. The oxygen was present in a sweep gas of helium. Under this condition abrupt output changes began to occur, followed eventually by thermocouple failure which was caused by oxidation of the wires.

Carroll and Reagan also stated that thermocouples in direct contact with graphite, while in an inert atmosphere at temperatures above about 900°C, showed a gradual negative drift. Breakage from brittle failure after several days exposure was reported. Hall and Spooner (1963) also mentioned high temperature incompatibility with graphite. Bennett and Hemphill (1963) observed considerable drift during tests in an atmosphere of helium with 100 ppm carbon monoxide. The temperature was 1750°C and test times were 50 to 75 hours. Similar results were observed during 216 hours at 1370°C.

Other experiments with thermocouples in insulators of several metal oxides were described by Hall and Spooner (1963), Farrow and Levitt (1964), Walker, Ewing, and Miller (1965, p. 816), Briggs and Johnston (1966), Hendricks and McElroy (1966), and Asamoto and Novak (1967). The work by Farrow and Levitt consisted of cycling tests of thermocouples with exposed junctions and with the wires in thorium dioxide insulators. These couples were in a carbon-contaminated atmosphere at 8 to 10 torr. The temperature ranged up to 1400°C, but measurements were made at 550°C. A positive drift of 2.5 per cent was observed during these tests. Tests at higher temperatures with lower pressures were also conducted, and reaction of the insulators with tungsten was observed above 2300°C, leading to thermocouple failure.

Walker, et al., reported reaction between W-26Re and aluminum oxide, which became extensive after about 140 hours at 1625°C in argon. During shorter periods, positive drift of 27°C and 37°C was found after 72 and 120 hours, respectively. Most of the drift was caused by changes in the W-26Re alloy.[13] The work of Hendricks and McElroy involved the use of aluminum oxide insulators of high purity and an exposure temperature of 1300°C in vacuum. Thermocouples in this environment for 352 hours showed no drift greater than ±5°C apparent change of temperature. An additional 1012 hours at 1450°C revealed no drift exceeding ±8°C. Briggs and Johnston tested two couples in beryllium oxide insulators for 610 hours at about 1920°C with drift of less than 1 per cent of that temperature.

The other investigations of metal oxide insulators utilized beryllium oxide and thorium dioxide. The exposure times were relatively short and in most instances voltage deviations were within experimental error or possibly caused by sources other than the oxides.

Differences in drift test results among the various investigators exist, usually because of the numerous factors which can affect drift and which are difficult to control. Results with insulated metal-sheathed thermocouples are even less clear-cut because of compatibility problems between the sheath, the oxide, and the wires. Also, thermal stresses become important, especially in the case of swaged or drawn cables. Finally, the electrical shunting of the thermocouple by the compacted oxide becomes a problem at high temperatures.

Fanciullo (1964, 1965) described the use of sheathed couples with aluminum oxide. Long-term drift tests were conducted using this insulation and with Nb-1Zr sheaths; an atmosphere of argon was present and the temperature was 1066°C. Provision was made to prevent air diffusion through the oxide from the cold end terminations of the sheaths; such diffusion had contaminated the thermocouple wires in earlier tests. The maximum drift observed after 10,000 hours was -0.55 per cent. Thermocouples of this type were also subjected to thermal cycling tests between approximately 200 and 700°C. No failures resulted during 200 cycles. Failures were encountered, however, when Type 316 stainless steel sheaths were used.

Studies of sheathed couples with beryllium oxide insulation were made by Hall and Spooner (1963), Clark (1964), Hendricks and McElroy (1966), Novak and Asamoto (1965, 1966, 1967), and Briggs and Johnston (1966). No chemical compatibility problems or drifts of any consequence were reported with the use of tantalum or W-26Re sheaths, although some thermal stress failures during cycling of tantalum-sheathed thermocouples were mentioned by Briggs and Johnston.

There was some evidence of reactivity and drift when molybdenum sheaths were used. Clark's test consisted of exposure of molybdenum-sheathed couples to temperatures of about 2320°C for 240 hours, including 60 cycles to room temperature and back. Negative drift of several per cent occurred during 240 hours total exposure, as measured at various calibration temperatures. Hall and Spooner described drift that was initially positive but then became negative.

However, the drift was small and within ±1 per cent at all temperatures during 100 hours above 2200°C. Here no large amount of thermal cycling was included in the test.

Black and Wilken (1963) used magnesium oxide as insulation and molybdenum and niobium as sheaths. Performance was satisfactory during 4 thermal cycles up to 1650°C. Lagedrost (1964) conducted a large number of cycling tests with the same type of insulation but with Nb-1Zr sheaths. Each thermal cycle consisted of a 1-minute temperature excursion between about 540 and 1540°C. Thermocouples with ungrounded junctions were subjected to 5000 cycles without failure. Frequent failures in less than 300 cycles occurred when the junctions were grounded, probably caused by thermal stresses. Clark (1964) reported calibrations of niobium-sheathed couples to 1650°C with good repeatability.

Clark also tested hafnium dioxide insulated thermocouples with molybdenum sheaths. Shunting effect at temperatures above 1600 to 1800°C was more pronounced than when beryllium oxide was used. An output decrease of 12 per cent was observed at 2300°C, compared with bare wire calibrations, caused by shunting.

Thorium dioxide applications were reviewed and studied by Clark (1964), Gross, et al. (1966), and Johnson (1966). Clark observed shunting errors in swaged couples which reached about −12 per cent at 2300°C and were noticeable in the 1600-1800°C range. Johnson, however, recommended thorium dioxide for use to at least 2200°C after reviewing earlier work. Gross and coworkers described insulated and coated probes which performed well in an oxidizing atmosphere at 14.5 torr.

Asamoto and Novak (1967) held a high-fired thorium dioxide insulated thermocouple at 2425°C for 148 hours in a vacuum furnace. The couple was sheathed with W-26Re alloy. Drift was within ±2 per cent under these conditions. These investigators also conducted tests with W-3Re vs W-25Re thermocouples, which can be expected to behave very similarly to W-5Re vs W-26Re in terms of stability and drift. Two such units, with W-26Re sheathing and vitrified thorium dioxide insulation, were tested to 2850°C, with shunting effects beginning at 2300°C and reaching 8 per cent of the output at 2850°C.

Villamayor (1968) performed thermal shock tests on various types of sheathed W-5Re vs W-26Re thermocouples in the 1000 to 2000°C range. When the temperature rate-of-change exceeded about 5°C/sec considerable changes were observed in the post-shock calibrations, compared with the initial calibrations. Prolonged heat treatment at 1500°C did not eliminate these changes.

Nuclear applications work has been discussed by Carroll and Reagan (1966), Bliss and Fanciullo (1966), Hancock (1967), Jeffs (1967), Villamayor (1968), and Novak and Asamoto (1968). The latter's irradiation test results generally confirmed earlier work concerning the drift due to transmutation, and the dependence of drift direction on temperature. For an accumulated exposure of 2.8×10^{20} n/cm^2 there was a negative drift of –5 per cent at about 480°C and +0.3 per cent at 1650°C. Work on this and similar thermocouples was reviewed by Heckelman and Kozar (1971). Their own work consisted of drift studies of irradiated and unirradiated W-3Re vs W-25Re thermocouples.

W-7Re vs W-26Re. This thermocouple was investigated by Lachman (1961), and a voltage-temperature curve, constructed from a table in this reference, is shown in Figure 5-7. The properties are very similar to those of W-5Re vs W-26Re.

Re-6W vs W-26Re, Re-9W vs W-26Re, Re-12W vs W-26Re, Re-15W vs W-26Re, W-26Re vs W-30Re. This set of high rhenium content alloys was studied by Kuhlman (1963, May; 1963, Sept.) during a thermocouple development program. Voltage-temperature curves are shown in Figure 5-8. The 12 and 15 per cent tungsten alloys were considered promising for thermocouple development; all of the high rhenium content alloys were reported to be strong and ductile. For the case of Re-12W vs W-26Re, a curve given for thermoelectric power S versus temperature showed values of S of approximately 5.8, 6.9, and 7.9 $\mu V/°C$ at 1600, 2000, and 2300°C, respectively.

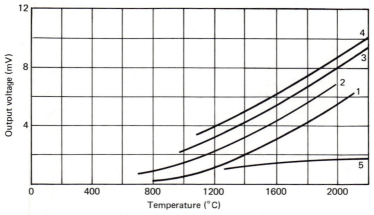

FIGURE 5-8 Tungsten-rhenium alloy thermocouples, III.   1. Re-6W vs W-26Re;   2. Re-9W vs W-26Re;   3. Re-12W vs W-26Re; 4. Re-15W vs W-26Re;   5. W-26Re vs W-30Re: Kuhlman (1963, May).

W-3Re vs Re, W-5Re vs Re. The voltage-temperature characteristics of these two thermocouples were constructed for Figure 5-9, based on separate curves of each metal referenced to tungsten from the work by Lachman and McGurty

(1962, p. 177). Walker, Ewing, and Miller (1965, p. 816) included these pairs with a set of 13 refractory metal thermocouples in studies of drift, and these couples proved to be superior in performance to the others. Such combinations as W vs W-26Re and W-5Re vs W-26Re were among those showing more drift than W-3Re vs Re and W-5Re vs Re.

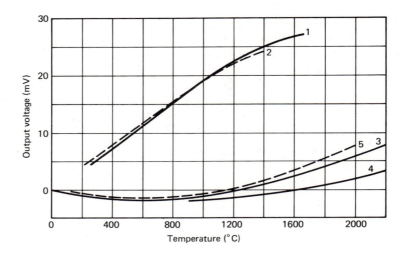

FIGURE 5-9   Rhenium paired with tungsten-rhenium   alloys. 1. W-3Re vs Re, Lachman and McGurty (1962, p. 177).   2. W-5Re vs Re, Lachman and McGurty (1962, p. 177).   3. Re vs W-25Re, Nadler and Kempter (1961).   4. Re vs W-26Re, Kuhlman (1963, May).   5. Re vs W-30Re, Sims, Gaines, and Jaffee (1959).

Thermocouples for the drift tests were sheathed in high purity aluminum oxide insulators and held at 1625°C in argon for 120 hours. Calibration changes of –1 and +4°C were observed for the W-3Re vs Re and the W-5Re vs Re thermocouples, respectively, measured at that temperature. The behavior of one wire batch indicated that a 4-hour heat treatment at 1625°C may be necessary in some cases to stabilize the output before use at or below that temperature.

Re vs W-25Re.[14]   Nadler and Kempter (1961) calibrated this couple during a study of thermocouple performance in carbon-contaminated atmospheres. Three calibrations were made in a graphite tube furnace through which there was a low flowrate of helium. The averaged data curve is given in Figure 5-9.

Re vs W-26Re.[15]   Kuhlman (1963, May; 1963, Sept.) included rhenium as well as a number of its alloys in a thermocouple study program, and the resulting voltage-temperature characteristic for this couple is shown in Figure 5-9. From other data by Kuhlman on thermoelectric power, the latter was found to be about 4.2, 6.1, and 7.4 $\mu V/°C$ at 1600, 2000, and 2300°C, respectively.

Walker, Ewing, and Miller (1965, p. 816) included this thermocouple with others in a drift investigation. The thermocouples were in high purity aluminum oxide insulators and in an atmosphere of argon at 1650°C. After 20 hours of exposure a positive drift of 23°C in the apparent temperature was recorded. After 72 and 120 hours the drifts were 31 and 43°C, respectively. These results showed considerably higher drift rates for Re vs W-26Re than for W-3Re vs Re and W-5Re vs Re. Other data clearly showed that the higher drift was due to the W-26Re alloy.

Re vs W-30Re.[16]   The voltage-temperature curve from Sims, Gaines, and Jaffee (1959) is given in Figure 5-9. Four calibration runs were made in a vacuum of 10^{-4} torr. The thermoelectric power, from the curve, is about 11.6 $\mu V/°C$ in the 1800-2000°C region.

W-3Re vs Mo, W-5Re vs Mo.   These thermocouples were evaluated by Walker, Ewing, and Miller (1965, p. 816) for drift in an argon atmosphere at 1625°C; the thermocouples were enclosed in aluminum oxide insulators. After 120 hours a drift of -2°C was found in the case of the 3 per cent alloy thermocouple and +5°C for the other couple. The calibration temperature for these data was also 1625°C. The drift results were considered to be favorable for exposures up to 150 hours. However, it was found that during longer periods molybdenum reaction with the aluminum oxide proceeded rapidly and led to failure of the wire.

Mo vs W-26Re.   A calibration was made by Kuhlman (1963, May), and also by Fanciullo (1964) for the region of lower temperatures. These results are plotted in Figure 5-10. From the results of Kuhlman, the thermoelectric power is approximately 16.2, 13.1, and 6.4 $\mu V/°C$ at 1200, 1600, and 2300°C, respectively. Walker, Ewing, and Miller (1965, p. 816) subjected this combination to a drift test at 1625°C in an atmosphere of argon. The thermocouples were sheathed in aluminum oxide insulators. After 72 hours a positive drift of 25°C was observed; after 120 hours the drift had increased to 39°C. Molybdenum was said to react rapidly with the aluminum oxide after about 150 hours at 1625°C, leading to failure of the wire.

Extensive work with sheathed thermocouples of this type was done by Fanciullo (1964, 1965). Grounded junction thermocouples were used, which were insulated with aluminum oxide and sheathed with Nb-1Zr. The cold ends of the thermocouple cables were either sealed or terminated in a vacuum chamber to prevent oxygen from diffusing through the insulation and reaching the hot zone. Tests were conducted at 1080°C in an atmosphere of helium. Initially, 7 thermocouples were within 3.9°C of indicating the same temperature. However, these thermocouples had an output of 35°C higher than the calibration curve determined earlier for Mo vs W-26Re. It was indicated that this deviation was the result of lot-to-lot variations in the wires, which could be eliminated by matching specific lots of molybdenum and W-26Re wires.

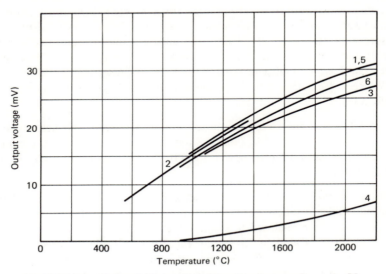

FIGURE 5-10   Other high temperature thermocouples.   1. Mo vs W-26Re, Kuhlman (1963, May).   2. Mo vs W-26Re, Fanciullo (1964).   3. Mo-10Re vs W-26Re;   4. Re-5Mo vs W-26Re;   5. Mo-10W vs W-26Re;   6. Mo-20W vs W-26Re: Kuhlman (1963, May).

A drift test of 10,000 hours at 1066°C was conducted and a maximum drift of 0.9 per cent in the direction of decreasing output was observed. The initial spread of 3.9°C had increased to 4.4°C after the test. The reactivity of molybdenum with aluminum oxide at 1625°C described by Walker, et al., was evidently insignificant at the lower temperature of 1066°C.

Mo-5Re vs W-26Re.   Zysk and Robertson (1971) considered this a combination worthy of further investigation, in a report on studies of a number of molybdenum-rhenium alloys. Voltage-temperature data given for Mo-5Re vs W-25Re were very similar to the Mo vs W-26Re curve of Kuhlman in Figure 5-10.

Mo-10Re vs W-26Re.   A calibration from work by Kuhlman (1963, May; 1963, Sept.) is shown in Figure 5-10. From the same source, the thermoelectric power is approximately 10.7, 8.3, and 6.4 μV/°C at 1600, 2000, and 2300°C, respectively. The molybdenum alloy was found to be similar to tungsten in strength but considerably more ductile.

Mo-50Re vs W-26Re. Cunningham and Goldthwaite (1966) included a section by C. B. Boyer and C. A. Alexander describing high pressure gas temperature measurements using this combination. Temperatures up to 1600°C were reached in atmospheres of argon or helium, with the presence of graphite. The molybdenum alloy wire was reported to have remained ductile at that temperature. No performance data were given.

Walker, Ewing, and Miller (1965, p. 816) conducted a drift test with thermocouples sheathed in aluminum oxide and exposed to a temperature of 1625°C in an argon atmosphere. After 20 hours a positive drift of 59°C was observed. After 120 hours the drift had increased to 83°C.

Re-5Mo vs W-26Re, Mo-10W vs W-26Re, Mo-20W vs W-26Re. These combinations were evaluated by Kuhlman (1963, May; 1963, Sept.) and the calibration data are included in Figure 5-10. Curves were also given by Kuhlman for the thermoelectric power S of these couples; for Re-5Mo vs W-26Re, S was about 5.2, 6.8, and 7.7 μV/°C at 1600, 2000, and 2300°C. In the case of Mo-20W vs W-26Re, S was about 12.2, 9.2, and 6.8 μV/°C at 1600, 2000, and 2300°C. Mo-10W appeared to have nearly the same characteristics as pure molybdenum, as shown in Figure 5-10 for Mo and Mo-10W referenced to W-26Re. The thermoelectric powers of both thermocouples were approximately 16.2, 13.1, and 9.5 μV/°C at 1200, 1600, and 2000°C, respectively.

Ta vs W-26Re. Davies (1960) evaluated this pair, making a calibration in vacuum. The voltage-temperature curve is unusual because of a very low output in the 0 to 1000°C region, followed by a rapid rise in output as the temperature is further increased. The high temperature portion of Davies' calibration is shown in Figure 5-11 for the useful output region of 1000 to 2800°C. It was observed that no cold junction compensation is required since the output is very low at lower temperatures. The thermoelectric power is about 13.6 μV/°C at 2000°C. Tantalum is so sensitive to contamination that it is not recommended for thermocouple use except in special circumstances.

THERMOCOUPLES WITH METALS OTHER THAN TUNGSTEN

Molybdenum and Molybdenum Alloys

Mo vs Nb. Calibration work was reported by Fanciullo (1964), Pletenetskii (1967), and Zysk and Robertson (1971). The results from Pletenetskii are shown in Figure 5-12. The niobium contained 0.57 per cent tantalum. The thermoelectric power S was found to be 14.7 μV/°C at 1000°C and 7.9 μV/°C at 1500°C. Aluminum oxide was recommended as the most suitable insulation material.

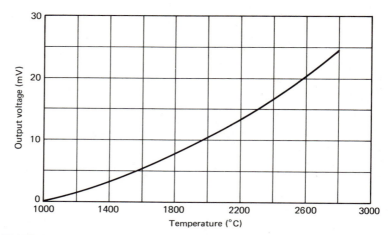

FIGURE 5-11 The Ta vs W-26Re thermocouple, from Davies (1960); output at temperatures below 1000°C is not shown.

A voltage-temperature curve from the results of Fanciullo is also included in Figure 5-12. The diameter of the niobium wire had a significant effect on the calibration, even though the wires were drawn from the same billet. The curve shown is for 1 mm diameter niobium. At 1040°C S is about 12 μV/°C.

The calibration by Zysk, et al., showed a decline in thermoelectric power at high temperatures, but not to the extent that the curve of Pletenetskii indicates. Values of S of 16 and 8 μV/°C were given for the temperatures of 1000 and 2000°C, respectively.

Both molybdenum and niobium are sensitive to impurities and are easily contaminated; this was discussed in the first section of this chapter. Insulation compatibility may also be a problem at temperatures near 1600°C, since Walker, Ewing, and Miller (1965, p. 816) found that molybdenum reacted with aluminum oxide after about 150 hours at 1625°C. Operation in clean, contaminant-free systems with high vacuum or very pure inert gases is a necessity for this combination. Niobium is ductile after a 10-minute anneal at 2200°C. Molybdenum becomes embrittled after high temperature exposure, but not nearly to the degree which tungsten does.

Bliss and Fanciullo (1965, 1966) conducted 10,000-hour drift tests at 1090°C with insulated Mo vs Nb couples sheathed in Nb-1Zr, and drift was within 0.5 per cent. These investigators, as well as Kuhlman (1966) and Villamayor (1968), considered Mo vs Nb attractive for nuclear applications because of the relatively low thermal neutron cross sections of both elements.

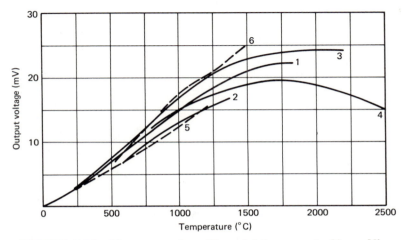

FIGURE 5-12   Thermocouples with molybdenum.   1. Mo vs Nb,
Pletenetskii (1967).   2. Mo vs Nb, Fanciullo (1964).   3. Mo vs
Re, Sims, Gaines, and Jaffee (1959).   4. Mo vs Ta, Morgan and
Danforth (1950).   5. Mo vs Ta, Roeser and Wensel (1941, p. 1293).
6. Mo vs Ti, Stadnyk and Samsonov (1964).

Mo vs Nb-1Zr.   Fanciullo (1964) described this thermocouple and attributed
earlier work on the combination to Holmes (1961). The use of niobium rather
than Nb-1Zr was recommended because of a tendency for drift which was observed when the latter alloy was used.

Mo vs Re.   Sims, Craighead, and coworkers (1956) reported a calibration made
up to about 2500°C. The average curve of calibration by Sims, Gaines, and
Jaffee (1959) is shown in Figure 5-12. These investigators also discussed early
work by J. Lachman, and data by G. Rouse of the National Bureau of Standards.
Other calibration data were given in the review by Caldwell (1962), by Lachman and McGurty (1962, p. 177), and by Danishevskii, et al. (1967). Using the
curve of Sims, Gaines, and Jaffee, the thermoelectric power was found to be
about 17 μV/°C at 1000°C. The data by Danishevskii, et al., led to about the
same value at that temperature.

    There is disagreement between some of the calibration curves, which is not
surprising in view of the sensitivity of molybdenum to small amounts of impurities and to contamination. Kuether (1960) conducted tests on the uniformity of
molybdenum and the resulting effect upon Mo vs Re thermocouples. Although
temperature calibration deviations as great as ±230°C were observed at 1950°C,

by proper selection of the molybdenum wires the deviations could be reduced to ±6°C.

Mo vs Mo-50Re.   A calibration was computed by Sims, Gaines, and Jaffee (1959) based on data for Mo vs Re and Re vs Mo-50Re. The voltage-temperature curve is shown in Figure 5-13. The thermoelectric power is near 18, 11, and 7 μV/°C at 800, 1600, and 2000°C, respectively. Stromberg and Stevens (1966), and Danishevskii, et al. (1967) also showed calibration data. The latter work included curves of thermoelectric power versus temperature.

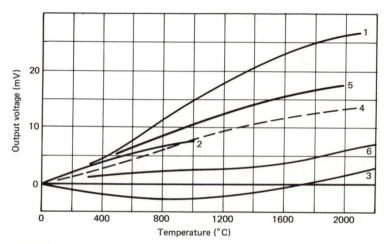

FIGURE 5-13.   Thermocouples with molybdenum, rhenium, and tantalum.   1. Mo vs Mo-50Re, Sims, Gaines, and Jaffee (1959). 2. Mo-30Re vs Re, Stromberg and Stevens (1966).   3. Re vs Mo-50Re, Sims, Gaines, and Jaffee (1959).   4. Mo-20Re vs Mo-40Re, Danishevskii, Ipatova, et al. (1967).   5. Mo-20Re vs Mo-50Re, Danishevskii, Ipatova, et al. (1967).   6. Ta vs Re, Lachman and McGurty (1962, p. 177).

Stromberg and Stevens found that there is an advantage in using this combination in high pressure apparatus because of low breakage where the wires are led through gasketed systems. Breakage was found to be much less than for Chromel-Alumel or platinum-rhodium alloy thermocouples.

High pressure calibrations were made relative to Chromel-Alumel as a comparison standard, and the relative pressure corrections were found to be small. Compared with Chromel-Alumel a pressure correction of about 7°C resulted for a pressure of 40 kB and a temperature of 1000°C. It was also found that strain or cold working tends to produce an apparent increase of pressure effect on temperature. Annealing at 1000°C for a short time after each pressure step helped to minimize this condition.

Mo vs Ta.   Morgan and Danforth (1950) included a calibration table for the 800-2500°C range, from which the curve in Figure 5-12 was plotted. Because of the voltage maximum near the relatively low temperature of 1750°C, the couple was not used for practical applications. A curve constructed from the tables of Roeser and Wensel (1941, p. 1293), which gives each element as referenced to platinum, is also shown for the 0 to 1200°C range. Rall (1960) described use of this pair for a nuclear application. One of the advantages for that situation was the low neutron cross sections of both elements.

Molybdenum and tantalum are both very sensitive to impurities and to contamination. The selection of either element for application with a second material of known stability generally results in a thermocouple that is difficult to use. The use of both elements together does not seem advisable under ordinary circumstances.

Mo vs Ti.   A voltage-temperature curve for this pair was given in a review by Stadnyk and Samsonov (1964), but no further information was included. The curve is shown in Figure 5-12.

Additional Molybdenum-Rhenium Alloys and Other Metals

Mo-30Re vs Re.   Stromberg and Stevens (1966) investigated this thermocouple as an alternative to Mo vs Mo-50Re. Their calibration table was used to plot the curve shown in Figure 5-13. The wires remained ductile after direct resistance welding of the junction, which was not the case for Mo vs Mo-50Re. This couple was considered applicable where extreme ruggedness and high temperature capability were required. Danishevskii, et al. (1967) evaluated each of these metals referenced to molybdenum during their investigation of Mo-Re alloys.

Re vs Mo-50Re.[17]   The calibration by Sims, Gaines, and Jaffee (1959) is given in Figure 5-13. The calibration work was done in a vacuum, and neutral or reducing atmospheres were also considered as permissible. Both metals remain ductile after recrystallization.

Walker, Ewing, and Miller (1965, p. 816) studied drift of thermocouples sheathed in high purity aluminum oxide insulators and held in argon at a temperature of 1625°C. After 20 hours a change in output equivalent to a decrease of 7°C was observed. After a total of 120 hours the drift had reversed direction and a net apparent temperature change of +10°C was observed.

Mo-5Re vs Mo-50Re, Mo-10Re vs Mo-50Re.   Zysk and Robertson (1971) discussed a study of molybdenum-rhenium alloys and considered the pair with

Mo-5Re as one of three combinations worthy of further investigation. The voltage-temperature curve is shown in Figure 5-14, from which the thermoelectric power S was found to be about 14 and 6 μV/°C at 1000 and 2000°C, respectively. The addition of 5 per cent rhenium improved the ductility of molybdenum for handling after high temperature exposure, but the alloy still did not have as good a bendability as Mo-50Re.

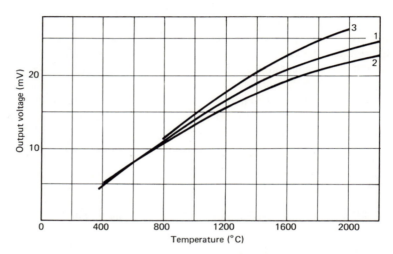

FIGURE 5-14   Additional thermocouples with molybdenum and rhenium.   1. Mo-5Re vs Mo-50Re;   2. Mo-10Re vs Mo-50Re: Zysk and Robertson (1971).   3. Mo vs Mo-50Re, Sims, Gaines, and Jaffee (1959).

The thermocouple with Mo-10Re, also from the paper by Zysk and Robertson, is shown in Figure 5-14. Values for S of 12 and 5 μV/°C at 1000 and 2000°C were found from the curve. Mo vs Mo-50Re is replotted in the same figure for comparison purposes.

Mo-20Re vs Mo-40Re.   Danishevskii, Ipatova, et al. (1967) discussed this thermocouple, and the calibration curve from their work is given in Figure 5-13. From other results in the same article, the thermoelectric power is about 7 μV/°C at 1000°C. The wires were said to remain ductile after annealing at 1800°C or higher, and the thermocouple was considered to be satisfactory for use in vacuum, inert gases, or hydrogen. Exposure to graphite resulted in embrittlement after 1 to 2 hours at 1500°C.

Mo-20Re vs Mo-50Re.   This was also investigated by Danishevskii and co-workers, with the resulting curve from the authors' Figure 5 shown in Figure

5-13. The thermoelectric power is 7.4 $\mu V/^\circ C$ at 1500°C. The wires remained ductile after exposure to temperatures of 1800°C or higher, and long term use in vacuum, inert gases, or hydrogen was suggested. Graphite embrittlement limits usage to 1 to 2 hours at 1500°C when that element is present.

Drift during the first 50 hours in vacuum at 1300 to 1900°C did not exceed 1 per cent. After 100 hours at fixed temperatures in this range the maximum observed drift, measured at 1300°C, was a slightly less than 2 per cent increase in voltage. The alloy with the higher molybdenum content was less stable than the Mo-50Re alloy.

Ta vs Re.   Sims, Craighead, et al. (1956) investigated this combination but considered the output to be too low for practical applications. They also observed instabilities from room temperature up to 2500°C. The calibration curve shown in Figure 5-13 was obtained from Lachman and McGurty (1962, p.177).

NONMETALS

Materials Properties and Thermocouple Design

Graphite requires some special comment because of the unusual properties of the material. Some of the properties of carbon or partially graphitized carbon are strongly affected by heat treatment and by the impurity content. The microstructure of the material changes during high temperature heating, and there is an increase in the degree of graphitization with corresponding changes in the thermoelectric power. Also, because of the initial impurities the thermoelectric power of a specimen can be either positive or negative relative to platinum. Finally, an initially pure graphite can be doped with an additive such as boron to deliberately change the thermoelectric power.

Because the thermoelectric power of graphite is dependent upon these factors it is difficult to fully describe the history and subsequent properties of a given sample. Only general descriptions are given here, but some of the references cited include such information as origin, heat treatment history, and impurity analyses. Reference to the literature is recommended when information on the properties of the material is required.

Other nonmetals are also affected by the relative amounts of the constituents, the impurities, and the method of preparation. Large differences in thermoelectric power are still common when comparing samples of a material from different sources. The situation is not usually so extreme as with graphite, however, because a fully stabilized microstructure can be obtained without lengthy high temperature heat treatment.

Nonmetals are generally brittle and lacking in tensile strength. Therefore, practical nonmetal thermocouples are usually large, heavy, composite rods or probes. The most common design consists of a rod of one material enclosed in a tube of the other material with a plug at the hot junction end where the two materials are in contact. The volume between the rod and the tube may be empty, packed with insulation, or filled with spacer rings. The cool end of the probe may be gas or water cooled, and in some cases matched extension wires are used from this location to a conventional reference junction. Depending upon the materials and the required length it may be possible to design rather small sized probes of this type. In the case of graphite, some unusual arrangements are possible, such as graphite yarn or filament thermocouples. Thin film couples using nonmetals have also been reported. The same general techniques used with other materials for thin films are employed.

Thermocouples with Graphite or Carbon

Au vs C.   A table by Fitterer (1933) showing nonmetallic thermocouples included gold versus carbon, attributed to Bartoli and Papasogli (1882). A curve is given in Figure 5-15 from data by Roeser and Wensel (1941, p. 1293).

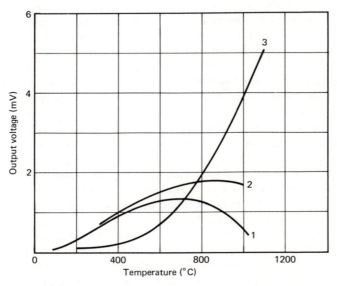

FIGURE 5-15   Thermocouples with carbon or graphite, I.   1. Au vs C;   2. Cu vs C;   3. C vs Ir: Roeser and Wensel (1941, p. 1293).

C(p) vs C(n).   Early work was done by Muraoka (1881). Bristol (1908) later described thermocouples of two differing types of graphite; Bidwell (1914), and

Moore (1915) reported similar work. All of these efforts and a considerable amount of later activity took advantage of the fact that natural or artificial graphites from different sources often have quite different thermoelectric properties.

Thielke and Shepard (1960) experimented with thermocouples of carbon and graphite in an effort to improve upon the results of Bristol and Bidwell, but concluded that this approach was unsatisfactory. They discussed a thermocouple which reached a maximum output at 1600°C but was not usable above about 1000°C because of the low thermoelectric power at the higher temperatures. Other workers have described thermocouples based on different forms of carbon and graphite, in some cases the output being fairly substantial. The contributing effects of natural or added impurities played an important part in some of the results obtained. Couples of this type were described by Franks (1962), Zuikov, Tsvetaev, and Bardyukov (1965), and Malyshev (1969). Samsonov and Kislyi (1967) also discussed graphite thermocouples in their review.

The effect of heat treatment on graphite was investigated by Thielke and Shepard, starting with an artificial graphite, and it was found that above 1000°C the thermoelectric power varied in a reproducible manner with baking temperature. A thermocouple formed of one rod baked to 2000°C and an identical rod baked to 3000°C produced an output of 25 mV at 1000°C. The thermoelectric power was 27.5 $\mu V/°C$ from 200 to 1000°C and was 3.8 $\mu V/°C$ at 2000°C. The latter temperature was the maximum limit for the thermocouple, since higher values would alter the characteristics of the rod previously baked at 2000°C.

Franks (1962) showed calibration curves for several commercial graphites with various impurities and described a useful application period of over 100 hours in a carbonaceous atmosphere at 2000°C. A useful period of 100 hours at 1700°C in vacuum, and maximum temperatures of 3000 and 2500°C for the carbonaceous atmosphere and for vacuum, respectively, were also given. The high temperature stability was within 1 per cent.

Another approach was the use of graphites of differing crystalline orientation. Ubbelohde, Blackman, and Dundas (1959) utilized pyrolytic graphite which was highly oriented with respect to the a-axis of the graphite lattice as the negative conductor, and a polycrystalline graphite as the positive member. The output was nearly linear between 1000 and 2300°C with a thermoelectric power of about 11 $\mu V/°C$. It was determined that defects in the carbon networks within the graphite were also in part responsible for the output, in addition to the difference due to crystalline orientation. A similar use of materials was described by Jamieson (1967) who patented a thermocouple composed of two different pyrolytic graphites.

Shepard, Pattin, and Westbrook (1956) studied the effect of small amounts of boron on the electrical properties of graphite and showed that the thermoelectric power S of pure graphite, which is -5 $\mu V/°C$ at room temperature relative to copper, is changed to +15 $\mu V/°C$ when about 1 per cent boron is added. A thermocouple with boron doped graphite was stable to at least 2600°C in neutral or reducing atmospheres. This type of thermocouple was described in more detail, and later work was also included, by Thielke and Shepard (1960), and Westbrook and Shepard (1960). It was shown that the doping of graphite with boron produced the maximum room temperature value of S when the boron content was 0.05 per cent. A much more heavily doped graphite was chosen for two reasons. First, small losses of boron from heavily doped graphite change the room temperature value of S very little. Second, the high temperature value of S remains relatively large for heavily doped graphite but decreases for low concentrations of boron.

A calibration curve of the boron doped thermocouple is shown in Figure 5-16 from Westbrook and Shepard. A number of applications for this couple were described by Thielke and Shepard. Long term operation at temperatures up to 2000°C was also discussed, and calibrations up to as high as 3000°C were made. The thermoelectric power was 35 $\mu V/°C$ at 2000°C. The output could be made reproducible to within 35°C from unit to unit, and accuracy was stated to be within ±10°C for individually calibrated couples. Thermocouple designs of the large coaxial probe type, small size couples, and flexible graphite yarn and filament couples were described.

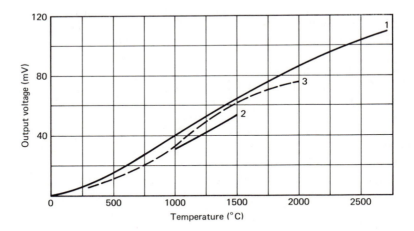

FIGURE 5-16   Thermocouples with carbon or graphite, II. The cold junctions are near 20°C.   1. C-1.0B vs C, Westbrook and Shepard (1960). (The exact boron content was not stated. The curve is believed to be typical of the 1 per cent boron doped type.)   2. C-1.0B vs C, Klein and Lepie (1964).   3. C-0.7B vs C, Klein and Lepie (1964).

Klein and Lepie (1964) and Klein, et al. (1964) investigated boron doped pyrolytic graphite versus undoped pyrolytic graphite, and observed results very similar to those of Thielke and Shepard for the behavior of S in terms of the amount of doping and the temperature. A voltage-temperature curve for a couple with 1 per cent boron content is included in Figure 5-16. These investigators preferred a 0.7 per cent boron content, however, because of its near linearity over a wide range and its high output. Also, the 1 per cent boron composition may exceed the solubility limit of boron in graphite. A curve for C-0.7B vs C is also shown in Figure 5-16. A graph of S as a function of temperature was given by Klein and Lepie for the 0.7 per cent boron doped couple, from which the values of 38, 54, and 36 $\mu V/°C$ at 500, 1000, and 2000°C, respectively, were obtained.

Heat treatment was found to have a pronounced effect on output and it was emphasized that the thermocouples must be heat treated at a temperature substantially above the operating temperature range in order to avoid instabilities. The C-0.7B vs C thermocouple was heat treated at 2900°C and was usable to about 2200°C. Above this temperature S was too low for accurate measurement applications. It was found that there was very little drift during periods of up to 1 hour at temperatures above 2000°C in a nitrogen atmosphere. Instabilities during calibration were not appreciable below 2000°C. Boron migration at the hot junction did not appear to be significant.

The C-0.7B vs C thermocouple can be used for a short time without prior heat treatment at temperatures above 2200°C, because in this condition S is still about 8.4 $\mu V/°C$ in that region. Details were not available on how much use could be obtained with as-received materials prior to occurrence of significant drift, and additional work on this type of approach was suggested.

A boron doped graphite thermocouple is commercially available (Anon., 1963), and the characteristics were summarized by Zysk and Robertson (1971). The unit provides stable long term operation to 2000°C in inert, carbonaceous, or reducing atmospheres, with an output of 80 mV at that temperature.

Cu vs C. The combination of copper-carbon was mentioned by Fitterer (1933) and was attributed to Weiss and Königsberger (1909). Moore (1915) used a copper versus carbon thermocouple while investigating the properties of carbon. A curve constructed from data by Roeser and Wensel (1941, p. 1293) is given in Figure 5-15.

C vs Ir. Carbon versus iridium from the work by Roeser and Wensel (1941, p. 1293), is shown in Figure 5-15. Two graphite-iridium thermocouples were described by Nadler and Kempter (1961). One variety of graphite used was positive,[18] the other negative,[19] with respect to iridium. Results by Bertodo (1963), with a graphite[20] which was positive with respect to iridium, indicated a thermoelectric power of 53 $\mu V/°C$ at 1600°C.

C vs Ni.   The voltage-temperature curve from data by Roeser and Wensel (1941, p. 1293) is plotted in Figure 5-17 for carbon vs nickel. According to Thielke and Shepard (1960), an early description of this thermocouple was given by an anonymous writer in 1918, but it was not until much later that a reference by Keinath (1935) was reported. Schulze (1939) reviewed this combination and stated that a usable life of over 300 hours can be obtained at 1200°C. It was indicated that the thermocouple is affected by drift, however, due to interaction of the two elements.

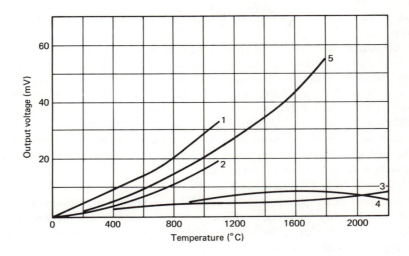

FIGURE 5-17   Thermocouples with carbon or graphite, III.   1. C vs Ni, Roeser and Wensel (1941, p. 1293).   2. C vs Pt, Roeser and Wensel (1941, p. 1293).   3. C vs Re, Nadler and Kempter (1961). 4. C vs Re, Salzano (1966).   5. W vs C, Franks (1962).

C vs Pt.   Early work on carbon-platinum was listed by Fitterer (1933) in a review section on nonmetal thermocouples; contributions were made by Bartoli and Papasogli (1882), Buchanan (1885), and La Rosa (1916). The voltage-temperature curve from work by Roeser and Wensel (1941, p. 1293) appears in Figure 5-17. Franks (1962) included curves of several graphites referenced to platinum.

C vs Re.[21]   Thermocouples of three different graphites versus rhenium were described by Nadler and Kempter (1961). One combination[22] is shown in Figure 5-17. The other graphites were negative with respect to rhenium. It was indicated that drift caused by diffusion of carbon into the rhenium would require in-place calibration just prior to use. Eshaya (1963) selected this pair because rhenium does not form carbides, and it was found that at least short term high temperature use was possible.

Salzano (1966) confirmed the general results of Eshaya by studies up to 2200°C in vacuum. A calibration curve from this work is shown in Figure 5-17; the maximum output of 8.5 mV was reached at about 1600°C in the case of a thermocouple which had been previously heat treated for 5 hours at 2200°C. Drift at an exposure temperature of 2200°C was observed that depended on the calibration temperature. Below 1600°C output increased with time whereas above that temperature the output decreased. The rate of drift decreased with time, and after 30 hours exposure at 2200°C it appeared that the drift was approaching zero, which was indicative of a stable condition.

On the basis of the drift results and other information it was suggested that an alloy of Re-11.7C would be stable in contact with graphite. Such a combination would be useful to just below the Re-C eutectic temperature of 2480°C, provided that the presence of the maximum in the voltage-temperature curve near 1600°C was acceptable from the measurements standpoint.

C vs W.[23]   Calibrations and other results for graphite with tungsten were given by Watson and Abrams (1928), Fitterer (1933), Holtby (1940, 1941), Losana (1940), Clark (1946), Franks (1962), and Bertodo (1963). A voltage-temperature curve from information on pure graphite and tungsten given by Franks is shown in Figure 5-17. Pure graphite is negative with respect to tungsten. The thermoelectric power is 32 μV/°C in the 1000°C region.

This thermocouple has been used to some degree in practical applications. Watson and Abrams described furnace temperature measurement and control for various uses and reported brief exposures at temperatures as high as 2400°C. Holtby and Clark were interested in liquid steel and cast iron applications. Franks did not recommend use in an oxidizing atmosphere, but indicated that exposure for over 100 hours in carbonaceous atmospheres to 1500°C or in vacuum to 1400°C was acceptable. One drawback for some applications was pointed out by Clark; this consisted of the sensitivity of the thermocouple output to moisture absorption by graphite. The effect is reversible and disappears when the moisture is baked out.

B$_4$C vs C.   A thermocouple of this composition was described by Ridgway (1939) for use in the 200 to 2400°C temperature range. The melting temperature of the boron carbide was about 2470°C. This couple was intended for use in reducing atmospheres, in the presence of metallic vapors, or for other extreme high temperature conditions. The voltage-temperature curve was nearly linear between 700 and 1700°C with a thermoelectric power of 340 μV/°C. An output of 590 mV was obtained at 2000°C, as shown in Figure 5-18. According to

Samsonov and Kislyi (1967, p. 6) the boron carbide electrode is difficult to make. Also, the characteristics of Ridgway's boron carbide were said to be more like those of an alloy of boron carbide with either boron or carbon, rather than those of pure B_4C.

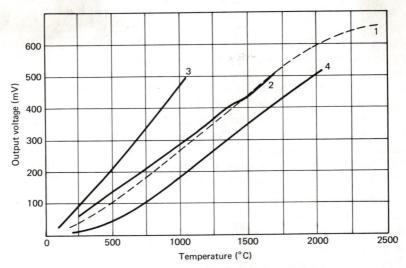

FIGURE 5-18   Thermocouples of several types with very high outputs. (In curves 1 and 2, the cold junction was near room temperature.)   1. B_4C vs C, Ridgway (1939).   2. C vs SiC, Fitterer (1933). 3. SiC(p) vs SiC(n), Bertodo (1963).   4. W-2(ThO_2) vs SiC, Bertodo (1963).

Cr_2O_3 vs C.[24]   Samsonov and Kislyi (1967, p. 8) prepared thermocouples of chromic oxide versus graphite and partially reduced some of the oxide during a sintering process in order to improve the electrical conductivity of the chromic oxide at lower temperatures. The low conductivity of the pure oxide was considered to be a major objection to this type of thermocouple. The couples consisted of graphite rods within chromic oxide tubes. The voltage-temperature characteristic is shown in Figure 5-19. The thermoelectric power averages 140 μV/$^\circ$C between 1200 and 1400°C.

This thermocouple was relatively drift-free during 20 to 25 hours at 1500°C in an oxidizing atmosphere. A drift exceeding 1.5 per cent occurred after longer periods of time. This was caused by oxidation of the partially reduced chromic oxide to its original condition. The possibility of doping the oxide with other additives was mentioned, and it was the opinion of the investigators that a fairly stable conductor would result.

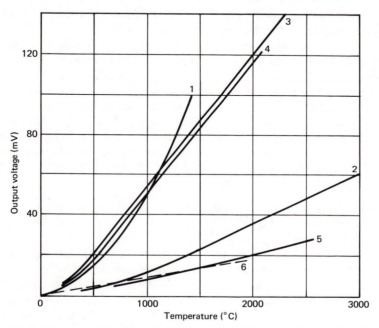

FIGURE 5-19 Thermocouples with graphite, carbides, and other materials. 1. Cr_2O_3 vs C, Samsonov and Kislyi (1967, p. 8) (Polarity was not specified.) 2. C vs NbC, Panasyuk, and Samsonov (1963). 3. C vs TiC, Samsonov and Kislyi (1967, p. 97). 4. C vs ZrB_2, Samsonov and Kislyi (1967, p. 97). 5. NbC vs ZrC, Panasyuk and Samsonov (1963). 6. ZrB_2 vs ZrC, Samsonov and Kislyi (1967, p. 97).

C^{25} **vs** $MoSi_2$. Bertodo (1963) evaluated this pair and developed a curve of thermoelectric power S versus temperature between 200 and 1600°C. Values of S from the curve were 48, 65, and 76 $\mu V/°C$ at 400, 800, and 1200°C, respectively. Oxidation tests showed that $MoSi_2$ was considerably less resistant than was silicon carbide in terms of weight loss.

C vs MoC.   Fitterer (1933) prepared rods of MoC and calibrated with respect to graphite from 300 to about 1060°C. The thermoelectric power was about 17.5 μV/°C in this interval.

C vs NbC.   Panasyuk and Samsonov (1963) briefly outlined the preparation of materials and the fabrication of coaxial rod-and-tube thermocouples. Calibrations in argon atmosphere were said to be accurate to within ±1.5 per cent. The resulting voltage-temperature curve is given in Figure 5-19. The thermoelectric power S averages 25 μV/°C between 1200 and 3000°C.

Drift tests were made in a gas mixture of carbon monoxide, hydrogen, and nitrogen. A drift of 3 per cent, as measured at 1000°C, resulted after 12 hours exposure at 2200°C. Other similar tests in hydrogen resulted in a 6 per cent drift after 12 hours, again at 2200°C and measured at 1000°C. However, the drift measured at 1400°C was only 1.0 per cent or less in the gas mixture and 1.7 per cent in hydrogen. The investigators suggested that NbC vs ZrC has some advantages over this combination, because of better high temperature stability. Samsonov and Kislyi (1967, p. 97) gave the upper temperature limit of C vs NbC at 2500°C. The thermocouple was said to be linear in the high temperature region with a value for S of 26 μV/°C.

C vs SiC.   The first use of graphite versus silicon carbide was reported by Fitterer (1933, 1936, 1941).[26] Thermocouples up to 2.5 meters in length were prepared using Globar silicon carbide rods within graphite tubes of unspecified origin. A 0.5-hour heat treatment at 1500 to 1600°C was recommended prior to calibration. Calibration data for selected temperature intervals were given up to 1700°C and a maximum operating temperature of at least 1700 to 1800°C was stated. The voltage-temperature curve is shown in Figure 5-18. The thermoelectric power S was given as 300 μV/°C over most of the temperature range. The calibration curves were within 1 per cent of an average curve for over 80 per cent of the thermocouples constructed. According to the 1941 reference over 350 such couples had been calibrated. Tests were also reported on some of these thermocouples. Repeated immersion in liquid steel and cycling experiments between room temperature and 1980°C produced no calibration changes greater than ±2.8°C. This type of couple and accessories for it were available commercially, Anon. (1937), and were used in the steel industry for several years.

Other calibration and evaluation work, using silicon carbide and carbon or graphite from different sources, was done by Losana (1940), Franks (1962), Bertodo (1963), and Samsonov and Kislyi (1967, p. 5). Voltage-temperature results varied considerably.

Franks indicated a useful life of 100 hours at 1750°C in a carbonaceous atmosphere, or 100 hours at 1200°C in vacuum. Use in an oxidizing atmosphere is very limited. Clark (1946) described steel industry applications and pointed out the sensitivity of the thermocouple to the presence of moisture. The output becomes low under this condition but can be restored by baking the thermocouple at a high temperature to dry it out.

Samsonov and Kislyi stated that large variations in characteristics which they and other observers had reported were due to the semiconductor properties of the silicon carbide, which make it sensitive to impurities, crystalline modification, and distribution of vacancies. A very dense, low porosity silicon carbide was used to obtain more consistent uniformity. This material, with graphite, had a value of S of 280 to 300 μV/°C above 400°C and about 62 μV/°C below that temperature.

TaC vs C. This thermocouple was mentioned by Keinath (1935); Samsonov and Kislyi (1967, p. 27) included tantalum carbide with other nonmetals considered suitable for thermocouple use.

C vs TiB$_2$. Bertodo (1963) presented a graph showing thermoelectric powers of 43, 60, and 67 μV/°C at 400, 800, and 1200°C, respectively. A commercial graphite[27] was used. The oxidation resistance of TiB$_2$ was rather low and silicon carbide was considered a better choice for an oxidizing atmosphere application.

C vs TiC. Bertodo also evaluated this combination. From his thermoelectric power vs temperature curve, S is 93, 103, and 110 μV/°C at 800, 1200, and 1600°C. The oxidation resistance of the carbide was less than one-third of that of silicon carbide at 1600°C.

Samsonov and Kislyi (1967, pp. 62, 97, 116) recommended this thermocouple for use in reducing or inert atmospheres, and for vacuum applications, at temperatures up to 2500°C. A voltage-temperature curve is shown in Figure 5-19. The thermoelectric power for the 600 to 2500°C interval was given as 65 μV/°C, a lower figure than the results of Bertodo. Uniformity of calibration from batch to batch of materials was within ±6 per cent, but by maintaining limitations on the titanium carbide composition a 2 per cent uniformity could be achieved. Most variations were due to changes in the voltage-temperature characteristic below 600°C, so the value of S at high temperatures was nearly independent of batch to batch differences. Within a given batch, calibration differences were within ±1.3 per cent.

Thermocouples were heat treated at 2000 to 2200°C for 2 hours and were then subjected to drift evaluation. Exposure cycles consisted of 10 to 15 hours at temperatures of 2000 to 2450°C in argon or hydrogen atmospheres, followed by recalibration. A negative drift was observed which was more pronounced at lower calibration temperatures than at high temperatures. After 25 hours in argon at 2200°C a drift of –1.6 per cent of the 1200°C calibration temperature was reported. In hydrogen the drift was –2.8 per cent, measured at 1200°C, after 45 hours at 2450°C. The drift was considerably higher at lower calibration temperatures after exposure to these same conditions.

Samsonov and Kislyi were of the opinion that drift in the 2000 to 2500°C range was less than 0.5 per cent in magnitude, since drift decreased as a function of calibration temperature at a given exposure. A practical application was also described for furnace temperature measurements. The temperature varied from 2150 to 2400°C and the atmosphere was carbon monoxide and nitrogen. The Russian designation for this thermocouple is TGKT.

C vs WC.   Fitterer (1933) prepared a rod of tungsten carbide and calibrated it with respect to carbon, obtaining a value of about 16 $\mu V/°C$ for the thermoelectric power in the 600-800°C region. Bertodo (1963) showed a larger figure of 49 $\mu V/°C$ at 800°C using a commercial graphite; this figure was taken from a curve of thermoelectric power versus temperature, which was given for the 200 to 1600°C range.

C vs ZrB_2.   The voltage-temperature curve in Figure 5-19, from Samsonov and Kislyi (1967, pp. 62, 97, 116) is very similar to that of C vs TiC. The thermoelectric power is 65 $\mu V/°C$ above 600°C. Graphite-zirconium diboride was recommended for temperature measurement of molten steel, cast iron, nonferrous metals, and some rare metals, for temperatures up to 1700-1800°C. It was also recommended for carburizing atmospheres to 2000°C. The uniformity from batch to batch of materials could be maintained within ±2 per cent and within one batch could be held within ±1.3 per cent, just as for C vs TiC which was described earlier. The behavior of calibration variations below 600°C and the decrease of drift at high calibration temperatures for a given exposure were also true for both C vs ZrB_2 and C vs TiC.

Drift tests were conducted following a 2-hour heat treatment at 2000 to 2200°C. After 25 hours at 2000°C in a hydrogen atmosphere, drifts of 5.2, 3.6, and 1.8 per cent were observed at calibration temperatures of 800, 1000, and 1200°C, respectively. The thermocouple was not recommended for thermal shock conditions because of consequent failure caused by thermal stresses. Failure caused by rapid immersion in liquid steel was cited as an example. The Russian designation for this couple is TGBTs.

Non-Graphite Combinations

NbC vs HfC.   This was described by Samsonov and Kislyi (1967, p. 62) as a promising combination for reducing or inert atmospheres, or for vacuum, at temperatures as high as 3000 to 3500°C.

NbC vs ZrC.   A voltage-temperature curve from the work of Panasyuk and Samsonov (1963) is included in Figure 5-19. Above 1100°C the curve is linear with a thermoelectric power of about 13 $\mu V/°C$. The thermocouples were constructed in the coaxial form of a center rod enclosed by a tube. Calibrations were made in an argon atmosphere and were reported to be accurate to within ±1.5 per cent.

Drift tests were conducted by making a calibration to 2000°C, exposing the thermocouple to a higher temperature for a period of time, and then recalibrating. In a mixed gas atmosphere of carbon monoxide, hydrogen, and nitrogen, a 12-hour exposure at 2200°C resulted in a -1 per cent drift at the calibration temperature of 1400°C. For 10-hour exposure temperatures of 2500 and 2600°C the corresponding drifts were -1 and -0.9 per cent, measured at 1400°C. Drift was somewhat higher in hydrogen: after 12 hours at 2200°C the calibration change was -1.5 per cent. However, after 10 hours each at 2500 and 2600°C the drifts were -0.9 and -0.5 per cent, respectively. In all cases, for a given exposure condition the drift decreased with increasing calibration temperature. The investigators were of the opinion that the drift measured at 2500 to 2700°C under these conditions would be as low as -0.5 to -0.7 per cent.

Samsonov and Kislyi (1967, p. 62) later suggested that this pair is promising for applications at temperatures up to 3000-3500°C. Both reducing and inert atmosphere use and also vacuum applications were considered favorably.

ZrB$_2$ vs ZrC.   Samsonov, Kislyi, and Panasyuk (1962) described investigations of these materials in detail. Rods with 0.5 meter lengths and 6 mm diameters were used. These nonmetals were said to be compatible with hydrogen, carbon monoxide, molten steels and iron, and with at least some nonferrous metals.

Samsonov and Kislyi (1967, pp. 97, 100) showed a calibration curve up to 2000°C, which was also listed as the upper limit for measurement work. The curve is shown in Figure 5-19. The calibration is linear above 600°C with a thermoelectric power of 9 μV/°C. Drift testing at 1800°C for 25 hours in hydrogen was preceded by a 2-hour heat treatment at 2000 to 2200°C. The test resulted in a 1.1 per cent decrease in output at calibration temperatures of 800, 1000, and 1200°C.

Moeller, Noland, and Rhodes (1968) reviewed applications of several types of thermocouples and stated that Hessinger and coworkers (1965) investigated this and other nonmetallic thermocouples for use in high temperature oxidizing atmospheres. This couple was described as showing promise for such applications. ZrB$_2$ maintained its shape and integrity in oxidation tests and was compatible with beryllium oxide and thorium dioxide. Tests were conducted up to 2000°C.

B$_4$C vs SiC.   A thermocouple of boron carbide vs silicon carbide was included in patent claims by Ridgway (1939). The output was said to be on the order of twice that of the B$_4$C vs C thermocouple. This would indicate that at 2000°C an output of over 1000 mV or 1 volt would result. No detailed information was given on this combination.

SiC(p) vs SiC(n).   Franks (1962) prepared thermocouples of this type by selecting high and low output silicon carbide tubes, and obtained a net thermoelectric power S of about 50 μV/°C. He was also successful in preparing small pieces of positively doped SiC, and using these he obtained a value of over 200 μV/°C for

S. Applications under various conditions were recommended including at least 100 hours use at 1700, 1750, and 1200°C in oxidizing atmospheres, carbonaceous atmospheres, or vacuum, respectively. A maximum temperature of 2200°C in a carbonaceous atmosphere was given.

Bertodo (1963) evaluated thermocouples of coaxial design with a rod enclosed in a tube. Positive silicon carbide was aluminum doped, and the n-type silicon carbide was nitrogen doped. Calibrations were made in still air to 1700°C and in flowing argon to 2200°C. A voltage-temperature curve is shown in Figure 5-18. The thermoelectric power *S* is about 500 μV/°C, which is higher than that of the C vs SiC thermocouple. From other data by Bertodo on separate specimens of the two materials relative to graphite, *S* was 518, 576, and 603 μV/°C at 800, 1200, and 1600°C, respectively. Large output variations from specimens of apparently identical batches were reported. The resistivity of the materials was high, being more than 1 ohm cm.

Shaffer (1963) described silicon carbide thermocouples from differing types of commercial materials. The output of the all silicon carbide combination was linear from 600 to 2200°C. Samsonov and Kislyi (1967, p. 7) also mentioned results with this thermocouple. The combination was recommended for lengthy work in oxidizing atmospheres at 1500-1750°C.

Ir vs SiC.   Gee (1963) described a probe design consisting of an iridium wire coated with a layer of metal oxide insulation, which was in turn coated with a layer of self-bonded silicon carbide. Various protective methods for enclosing the thermocouple to complete the probe assembly were also discussed. The assembled probe was said to be capable of operating at temperatures as high as 2316°C in an oxidizing atmosphere, with an accuracy of ±1 per cent during the first 100 hours and on the order of ±2 per cent during the following 1000 hours. An output near 20 mV at 1927°C was mentioned.

W-2(ThO$_2$) vs SiC.   Bertodo (1963) developed this thermocouple for use in the combustion chamber and afterburner of turbojet engines. The thoriated tungsten was chosen because of the resulting high output with silicon carbide, combined with a low sensitivity to composition in the 2.0 to 2.5 per cent range of thorium dioxide. A voltage-temperature curve is plotted in Figure 5-18. From the curve, the thermoelectric power *S* is near 320 μV/°C in the 800 to 2000°C interval. Somewhat different values of *S* resulted from other data by Bertodo, apparently for different specimens.

The calibration and drift test of a concentric type thermocouple was described. This consisted of a thoriated tungsten rod enclosed in a tubular sheath of silicon carbide. Calibrations between 20 and 2200°C in still air or in flowing argon could be repeated within ±0.5 per cent. Three thermocouples exposed in air at 1700°C and at 2000°C showed stability within ±0.3 per cent for short periods of time, and drift within ±0.5 per cent during 25 hours of exposure.

Two factors determined the amount of drift. The first was the preferential oxidation of tungsten which resulted in enrichment of the thorium dioxide. The second factor was a migration of minor constituents in both conductors.

Thermocouples without Carbon Compounds

Cr_2O_3 vs Cr_2O_3-1 mole(TiO_2).[28]   Fischer and Lorenz (1958) studied this unusual combination and developed voltage-temperature curves and tables based on their calibration work. The curve for an argon atmosphere is included in Figure 5-20. It should be noted that the "cold" junction reference is at 300°C. Oxide resistivity becomes high at lower temperatures, so platinum extension wires were used from the 300°C reference junction into cooler regions. The thermoelectric power was 1250 μV/°C at 1000°C.

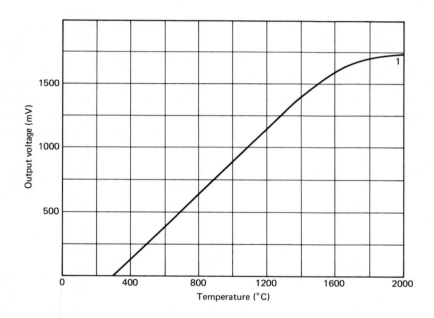

FIGURE 5-20   Thermocouple using chromium oxides. (The reference junction temperature was 300°C.)   1. Cr_2O_3 vs Cr_2O_3-1 mole (TiO_2), Fischer and Lorenz (1958).

Pt vs Cr_2O_3.   Vonkeman (1970) developed this combination as a simplified version of the Cr_2O_3 vs Cr_2O_3-1mole(TiO_2) thermocouple for use in a calorimeter. The unit consisted of a fused silicon dioxide tube containing a hollow cylinder of chromium oxide, through which an insulated platinum wire led to the closed hot junction end of the tube. The reference junction end[29] was held near 360°C because the high electrical resistivity of the chromium oxide did not permit operation at much lower temperatures. No calibration curve was given, but from results for thermoelectric power, average values of S were 617 and 613 μV/°C for the 419-776 and 776-961°C intervals, respectively. S was nearly constant from about 400°C to 1200-1300°C. Accuracies of 0.05 and 0.2°C at 419 and 960°C, respectively, were given for results from melting point experiments. A repeatability on the order of 0.01°C was stated for melting point measurements.

Cu vs MoS$_2$.[30]   Dutta and Bhattacharya (1965) studied a thermocouple of natural molybdenite with copper. The thermoelectric power was given as 160 μV/°C over the 0 to 250°C range. The molybdenite was in the form of a thin flake, with a 6 cm length.

Mo vs MoSi$_2$.   This combination was developed by Kröckel (1960) for applications in the glass making industry. The calibration curve from this reference is shown in Figure 5-21. The thermoelectric power is 16 μV/°C in the interval between about 600 to at least 1800°C. The unit was constructed in the form of a molybdenum wire, insulated with zirconium dioxide and held within a tube of MoSi$_2$. The wire joined the nonmetal in a plug of that material at the hot junction end of the tube. There was considerable drift during the first few hours of use with this arrangement as the molybdenum and the MoSi$_2$ interacted, but the output later became stable.

MoSi$_2$ vs WSi$_2$.[31]   Panasyuk and Samsonov (1963) mentioned this thermocouple in reviewing their earlier work, and considered it to be useful in oxidizing atmospheres up to 1700°C. With an outer jacket of ZrB$_2$ it was also usable in molten ferrous and nonferrous metals.

Samsonov and Kislyi (1967, pp. 62, 97, 113) gave the calibration shown in Figure 5-21. The thermoelectric power is 22 μV/°C in the 400-800°C range, and less at higher temperatures for the particular curve shown here. The heat treatment temperature used during fabrication as well as batch variations of the materials compositions affected the calibration. An application for furnace temperature measurements was described and caution was recommended to avoid thermal shock. The Russian designation for this couple is TMSV.

FOOTNOTES

1. Also called columbium in some branches of industry and engineering.
2. All references are located at the end of the book.

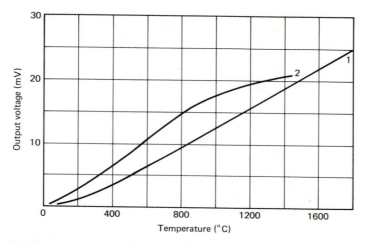

FIGURE 5-21   Thermocouples with molybdenum disilicide.   1. Mo vs MoSi$_2$, Kröckel (1960).   2. MoSi$_2$, vs WSi$_2$, Samsonov and Kislyi (1967, p. 99).

3. In the nomenclature of Lachman this instability is negative. As used here, positive instability indicates an increasing thermocouple voltage when the thermocouple is held at constant temperature.
4. In the nomenclature of Lachman and Kuether this instability would be negative.
5. From Figure 3 in the work by Danishevskii. Other data in this reference give slightly different results.
6. The original publication date was 1961.
7. Information on this couple became available shortly after publication of data on the very similar W-5 Re vs W-26Re pair. Some background information on these types is given under W-5Re vs W-26Re.
8. Engelhard Minerals and Chemicals Corp.
9. This was wire which had not been given the usual heat treatment supplied by the manufacturer prior to shipping.
10. The W-26Re alloy has been used on occasion in place of W-25Re. The actual compositions of these alloys may be even closer together than these nominal values indicate.
11. There are some minor differences between the tables in the two papers published by Lachman in 1961.

12. The shapes of the drift curves are quite different for the two types of thermocouples; because of this the noble metal type shows considerably less drift until 7000-8000 hours have passed.
13. These results conflict with some of the other data quoted here and may indicate that oxide insulation purity is critical for determining alloy reactivity.
14. Rhenium is positive above 1250°C.
15. Rhenium is positive above 1650°C.
16. Rhenium is positive above 1200°C.
17. Rhenium is positive above 1750°C.
18. National Carbon Co., CCU graphite.
19. Graphite Specialties Corp., Graph-i-tite "G."
20. Morgan Crucible Co. EY4 graphite.
21. Polarity was not specified in some of the work described here and generally depends upon the type of graphite used.
22. National Carbon Co., CCU graphite.
23. Carbon is positive; some graphites are negative with respect to tungsten.
24. Polarity was not specified.
25. Morgan Crucible Co., EY4 graphite.
26. Fitterer initially referred to the positive material as carbon, but in the 1941 reference it was described as a graphite element which was thoroughly graphitized after extrusion to the proper size.
27. Morgan Crucible Co., EY4 graphite.
28. One mole per cent of titanium dioxide was used.
29. The complete system was more complex than is indicated here, but this description is sufficient for discussing the Pt vs Cr_2O_3 thermocouple.
30. Polarity was not specified; the materials were listed in the order shown.
31. Polarity is uncertain. Notation in the reference by Samsonov and Kislyi quoted in this section indicates that $MoSi_2$ is positive. Data from tables of thermoelectric powers of materials from the same reference show that $MoSi_2$ would be negative.

Appendices

A. Summary of Thermocouple Types and Characteristics

A condensed list of the thermocouples described in the text is given here. Those combinations for which little or no information is available are not included. Thermocouples that have been used to a considerable degree in the United States are designated by a solid circle ●. Practice in other nations differs and does not always follow the usage indicated here. The group of specially designated couples is a relatively large one and there have been suggestions for standardizing thermocouple types to a smaller number for everyday requirements.

Wherever possible, some of the important thermocouple characteristics are included with each entry in the list. Characteristics such as thermocouple range and type of atmosphere refer to typical application conditions rather than extreme limits.

NOBLE METAL THERMOCOUPLES

Elemental and Binary Platinum Alloys

Type	Characteristics
Au vs Pt	Extremely stable. The voltage- temperature curve is highly reproducible over a range of 0-600°C.
Pt-10Ir vs Pt, Pt-15Ir vs Pt, Pt-20Ir vs Pt, Ir vs Pt	More susceptible to oxidation than comparable Pt-Rh couples, but can be used in air to 1600°C.
Ir vs Pt-20Ir, Ir vs Pt-30Ir	Low sensitivity. Susceptible to oxidation. Usable to 1900-2000°C.
Pt vs Pd	0 to 1300°C in oxidizing atmospheres.
● Pt-10Rh vs Pt	0 to 1500°C range for precision applications. For nonreducing atmospheres. Much data available on performance. Used in primary and secondary standard thermometers.
● Pt-13Rh vs Pt	Very similar to Pt-10Rh vs Pt, but has no particular advantages.
Pt-13Rh vs Pt-0.5Rh, Pt-13Rh vs Pt-1Rh	Somewhat improved high temperature drift compared with Pt-10Rh vs Pt.
Pt-20Rh vs Pt-5Rh, ● Pt-30Rh vs Pt-6Rh	Superior to Pt-10Rh vs Pt at high temperatures. 200 to 1700°C range. For nonreducing atmospheres.

226

Elemental and Binary Platinum Alloys (continued)

Pt-40Rh vs Pt-20Rh	For 1500 to 1800°C in nonreducing atmospheres.
Rh vs Pt	To 1500°C in nonreducing atmospheres.
Rh vs Pt-10Rh	For 600 to 1500°C in nonreducing atmospheres.
Rh vs Pt-20Rh, Rh vs Pt-30Rh	For use to 1800°C. Low sensitivity below 1000°C.
Pt-6Ru vs Pt	0 to 1600°C in inert atmospheres. Proposed for nuclear applications.
Pt-10Ir vs Au-40Pd	Pallador I. For 0 to 1000°C in air, and with high sensitivity.
Pt-10Rh vs Au-40Pd	Similar to Pt-10Ir vs Au-40Pd.
Pt-10Ir vs Pd, Pt-15Ir vs Pd	For oxidizing atmospheres, –60 to +1400°C. Thermal stress problems have been reported.
Pt-20Rh vs Pd	Range of at least 0 to 1300°C.
Ir vs Pt-10Rh	Subject to drift at 1380°C.

Multicomponent Systems Including Platinum

Type	Characteristics
Pd-12.5Pt vs Au-46Pd	Pallador II. For nonreducing atmospheres in the 0 to 1200°C range.
Pd-14Pt-3Au vs Au-35Pd	Platinel I. For oxidizing atmospheres in the 0 to 1200°C range.
● Pd-31Pt-14Au vs Au-35Pd	Platinel II. Improved strength. Output similar to that of the Type K couple.
Pt-5Rh vs Au-46Pd-2Pt	Pallaplat. 0 to 1200°C, used in many corrosive media.
Pt-20Rh-10Ir vs Pt-20Rh, Pt-20Rh-5Ru vs Pt-20Rh, Pt-30Rh-10Pd vs Pt-20Rh	Calibrated to about 1800°C, but subject to drift.

Elemental and Alloy Systems Excluding Platinum

Type	Characteristics
Ag vs Pd	Secondary laboratory standard for the 0 to 600°C range.
Rh vs Pd	To 1300°C in oxidizing atmospheres.
Ir-10Rh vs Ir, Ir-25Rh vs Ir, ● Ir-40Rh vs Ir, ● Ir-50Rh vs Ir, ● Rh-40Ir vs Ir, Rh-25Ir vs Ir	For vacuum, inert, and mildly oxidizing atmospheres to above 2000°C. Wire embrittlement at high temperatures. Low outputs.
Rh-10Ir vs Ir	To 1900°C. Low output.
Ir-50Rh vs Ir-10Rh	Low sensitivity, but useful above 2000°C.
Ir vs Ir-10Ru	Calibrated to 1800°C. Low output. Oxidizes in air. Wire embrittlement.

Elemental and Alloy Systems Excluding Platinum (continued)

Ir-10Rh vs Ir-10Ru, Rh-40Ir vs Ir-10Ru	High temperature oxidation in air, also wire embrittlement

THERMOCOUPLES OF MIXED NOBLE AND BASE METALS

Elemental and Binary Systems

With Platinum

Type	Characteristics
Pt vs Cu-0.0016at.In	Suggested for cryogenic use: 4.2 to 20 K. $S = 10 \, \mu V/K$ at 10 K.
Pt vs Au-0.02at.Fe	For 4.2 to 20 K. $S = 15 \, \mu V/K$ at 10-15 K.
Pt-5Mo vs Pt, Pt-5Mo vs Pt-0.1Mo, Pt-5Mo vs Pt-1Mo	Inert atmospheres from 0 to 1600°C. Nuclear applications.
Mo vs Pt	Similar to above Pt-Mo couples.
Mo vs Pt-10Rh	Calibrated from 400 to 1500°C.
Pt-1Re vs Pt, Pt-2Re vs Pt, Pt-8Re vs Pt	Inert atmospheres to 1400°C. Pt-8Re embrittles after heating.
Re vs Pt	100 to 1600°C in inert atmospheres.
Pt-8Re vs Rh	Range of 400 to 1600°C, with rapid drift in oxidizing atmospheres.
Te vs Pt	High output, but shows rapid drift above 90°C.
V vs Pt	Vacuum applications, 0 to 1300°C.

Without Platinum

Type	Characteristics
Ag vs Al	Thin film couple for the 4.2 to 300 K range. Linear above 20 K. $S = 3.1 \, \mu V/K$.
Ag vs Bi	Thermopile radiometers.
Ag vs Ni	Thin film applications.
Ag vs Cu-43Ni	0 to 600°C range. For higher temperatures than Cu vs Cu-43Ni.
Au vs Bi, Au vs Ni	Thin film applications.
Ag-0.37at.Au vs Au-2.1at.Co	4.2 to 300 K. Gold-cobalt is unstable above room temperature.
● Ag-0.37at.Au vs Au-(0.02-0.03)at.Fe	Range of 1 to 40 K. $S = 14 \, \mu V/K$ near 8 K.
Ag-0.37at.Au vs Cu-43Ni, Ag-28at.Au vs Cu-43Ni	Superior to Cu vs Cu-43Ni for the 4.2 to 300 K range when low thermal conductance is required.
Cu vs Au-2.1at.Co	4.2 to 300 K. Gold-cobalt is unstable above room temperature.
● Cu vs Au-(0.02-0.03)at.Fe, ● Cu vs Au-0.07at.Fe	Range of 1 to 40 K.

Elemental and Binary Systems Without Platinum (continued)

Nb vs Au-(0.02-0.03)at.Fe	Special applications below 9 K where Nb is superconductive.
Ni-10Cr vs Au-2.1at.Co	4.2 to 300 K. Gold-cobalt is unstable above room temperature.
● Ni-10Cr vs Au-(0.02-0.03)at.Fe, ● Ni-10Cr vs Au-0.07at.Fe	Range of 1 to 300 K. $S = 9\ \mu V/K$ at 1 K.
Ni-10Cr vs Au-(9-10)Ni	High output. Some evidence of instabilities in the 400-600°C region.
Re vs Ir	Very nonlinear output curve.
Re-30Ir vs Ir	For vacuum, inert gases, or slightly oxidizing atmospheres, from 600 to 2000°C.
Re-40Ir vs Ir	Similar to Re-30Ir vs Ir, but more sensitive to oxidation.
Re-30Ir vs Ir-20Re	Vacuum or inert atmospheres. Calibrated to 2300°C.
W vs Ir	To 2100°C in vacuum or inert atmospheres. Wire embrittlement after heating.
W vs Rh-40Ir	Calibrated to 1900°C in inert atmosphere.
Ni-10Cr vs Pd	0 to 1200°C.
W vs Rh	Calibrated to 1950°C.
Rh-8Re vs Rh	Calibrated to 1900°C. No recent reports on characteristics.

W-0.5at.Os vs W-(25-26)Re,
  W-1at.Os vs W-(25-26)Re,
  W-3at.Os vs W-26Re
	0 to 2000°C range in inert atmospheres for nuclear applications.

Multicomponent Systems

Type	Characteristics
Pt-5Rh-4.5Re vs Pt	0 to 1600°C. Superior to Pt-8Re vs Pt, but sensitive to oxidation at high temperatures.
Pt-20Rh-10W vs Pt-20Rh, Pt-20Rh-10W vs Pt-30Rh	Range of 0 to 1800°C, but drifts during air exposure.
Te-32Ag-27Cu-7Se-1S vs (Ag$_2$S)-50(Ag$_2$Se)	Infrared detector. High sensitivity: $S = 150$ to 500 $\mu V/^{\circ}$C.
Au vs (Ag$_2$S)-50(Ag$_2$Se)	Infrared detector. High sensitivity: $S = 400$ to 490 $\mu V/^{\circ}$C.

BASE METAL THERMOCOUPLES FOR
LOW AND MODERATE TEMPERATURES

Thermocouples Containing Copper

Type	Characteristics
Cu vs Fe, Cu vs Ni	Thin film applications.
Cu vs Cu-1.78Ni	Extension wires for W-5Re vs W-20Re.

Thermocouples Containing Copper (continued)

Cu vs Cu-3Ni	Extension wires for Pt-10Rh vs Pt and other combinations.
● Cu vs Cu-43Ni	Copper-constantan. For the −190 to +370°C range. Exceptionally good repeatability can be attained.
Cu vs Cu-0.005Sn	2 to 30 K.
Cu vs Fe-(17-19)Cr-(9-13)Ni-(0-2)Mn	Cu vs Type 347 stainless steel extension wires for Ir-40Rh vs Ir.
Fe vs (82-86)Cu-(4-15)Mn-(2-12)Ni-Fe	Iron-Manganin for W-5Re vs W-20Re extension wires.
● Fe vs Cu-43Ni	Iron-constantan. For the −190 to +760°C range in reducing, oxidizing, or inert atmospheres.
Zr vs Cu-43Ni	Special nuclear application.
Cu-3Ni vs Ni-3Mn-2Al-1Si	Extension wires for use with W-5Re vs W-26Re.
Cu-12.3Ni vs Cu-20Ni	Extension wires for use with Pt-10Rh vs Pt.
Cu-1.25Ni-0.74Mn-0.01C vs Ni-45Cu	Extension wires for Platinel I and Platinel II.
● Ni-10Cr vs Cu-43Ni	Chromel vs constantan. Range of 0 to 870°C. Not for reducing atmospheres. High output.
Ni-25Fe-11Cr vs Cu-45Ni	Chromel X vs Copel. Used for temperatures below 200°C.
Ni-16Mo vs Ni-50Cu	Range of 0 to 800°C. High output.
Ni-(15-25)Cr vs (96-99)Cu-(0.5-3.0)Ni-(0-1.5)Mn	Extension wires for W-3Re vs W-25Re.

Thermocouples Without Copper

Type	Characteristics
● Sb vs Bi	For radiometers, usually as thin film arrays.
Te vs Bi	For radiometers.
Sb-48.3Cd vs Bi, Sb vs Bi-10Sb, Sb-25Cd vs Bi-10Sb	Sensitive fine wire thermocouples and thermopiles.
Te-(0.4-0.5)Bi vs Bi-10Sb, Te-1.5S vs Bi-10Sb	Thermopile applications.
Bi-5Sn vs Bi, Bi-10Sn vs Bi, Bi-5Sn vs Bi-3Sb	For radiometers.
Ge(p) vs Ge(n)	Thin film couple with high sensitivity. To 100-200°C.
Fe vs Ni	Wire or thin film couples for special applications.
Mo vs Ni	Superior to Pt-10Rh vs Pt when direct contact with Ni is required.
Ni-(9-10)Cr vs Ni, Ni-14.7Cr vs Ni, Ni-16Cr vs Ni	Chromel-Alumel is often preferable to these.
Ni-34Fe vs Ni	0 to 1000°C.

Thermocouples Without Copper (continued)

Ni-18Mo vs Ni	0 to 1200°C. High output, for reducing or inert atmospheres.
Co-20Cr vs Co-5Fe	Range of 0 to 900°C.
Co-23.5Fe vs Co-17Fe	Extension wires for W-5Re vs W-26Re.
(95-97.5)Ni-(1.5-2.2)Cr-(0-1.2)W-(0.7-1.1)Al-(0.3-1.0)Si vs Ni-(3-5)W	Extension wires for W-3Re vs W-25Re.
• Ni-10Cr vs Ni-xMn-yAl-zSi	Chromel-Alumel and similar alloy (Type K) thermocouples for the 0 to 1260°C range. Not for reducing atmospheres.
Ni-9.5Cr-1.5Si vs Ni-4.25Si-0.25Mg	Improved performance compared with Type K couples.
Ni-22Fe-15Cr vs Ni-4.3Si	Extension wires for Pt-15Ir vs Pd.
• Ni-20Cr-1Nb-1Si vs Ni-3Si	Geminol. −18 to +1260°C. Improved reducing atmosphere performance compared with Type K couples.
Fe-(19-21)Cr-(10-12)Ni-(0-2)Mn-(0-1)Si vs Ni-3Mn-2Al-1Si	Type 308 stainless steel vs Alumel extension wires for W-3Re vs W-25Re.

BASE METAL AND NONMETAL THERMOCOUPLES FOR MODERATE AND HIGH TEMPERATURES

Thermocouples With Tungsten and Tungsten Alloys

Type	Characteristics
W-5Mo vs W, W vs W-15Mo, W vs W-25Mo, W vs W-35Mo, W vs W-50Mo	Not generally recommended.
W vs Mo	1300 to 2400°C range. Difficult to use because of lack of uniformity. Sensitive to contamination.
W vs Nb	100 to 1772°C in vacuum or inert atmospheres.
• W vs W-26Re	W with W-Re alloys of 26% Re or less are used from 0 to 2600°C in inert gases, H_2, or vacuum. Wire embrittlement after heating.
W vs Re	0 to 2000°C in inert atmospheres, H_2, or vacuum. Tungsten wire embrittlement after heating.
W vs Ta	Not generally recommended.
W-3Re vs W-15Re, W-3Re vs W-20Re, W-5Re vs W-20Re, W-10Re vs W-20Re W-15Re vs W-20Re	These thermocouples are in use outside the United States.
• W-3Re vs W-25Re, • W-5Re vs W-26Re	Superior to W vs W-26Re. 0 to 2800°C range in inert gases and H_2; also vacuum for short periods. Wire embrittlement after heating.
W-3Re vs Re, W-5Re vs Re	Improved compatibility with Al_2O_3.

Thermocouples With Tungsten and Tungsten Alloys (continued)

Re vs W-25Re, Re vs W-26Re	Very nonlinear.
W-3Re vs Mo, W-5Re vs Mo	Not generally recommended.
Mo vs W-26Re	Not generally recommended. However, specially prepared couples show low long term drift.
Mo-10Re vs W-26Re, Mo-50Re vs W-26Re	Mo-Re alloys are more ductile than tungsten after heating.
Ta vs W-26Re	1000 to 2800°C in high vacuum.

Thermocouples With Metals Other Than Tungsten

Type	Characteristics
Mo vs Nb	Not generally recommended.
Mo vs Re	Calibrations to 2000°C, with some lack of agreement.
Mo vs Mo-50Re	0 to 2000°C range. Problems may occur using Mo.
Mo vs Ta	Not generally recommended.
Mo-30Re vs Re, Re vs Mo-50Re	May be useful in special applications.
Mo-5Re vs Mo-50Re, Mo-10Re vs Mo-50Re, Mo-20Re vs Mo-40Re, Mo-20Re vs Mo-50Re	Some of these show promise for further investigation.

Nonmetals

Type	Characteristics
C(p) vs C(n)	Natural or artificial graphites for special applications.
C vs Re	High temperatures can be reached, but drift occurs.
C vs W	Nonoxidizing atmospheres, 0 to 2000°C.
C vs NbC	Nonoxidizing atmospheres, 1400 to 2500°C.
C vs SiC	Range of 100 to 1800°C with high sensitivity. Limited use in oxidizing atmospheres.
C vs TiC	Up to 2500°C in nonoxidizing atmospheres.
C vs ZrB$_2$	Up to 1800°C. Sensitive to thermal shock.
NbC vs ZrC	Nonoxidizing atmospheres to at least 2600°C.
ZrB$_2$ vs ZrC	0 to 2000°C range.
SiC(p) vs SiC(n)	High sensitivity: $S \geqslant 500\ \mu V/^\circ C$. To 1750°C in oxidizing atmospheres, higher in other gases.
W-2(ThO$_2$) vs SiC	20 to 2200°C in still air with W protected.
Cr$_2$O$_3$ vs Cr$_2$O$_3$-1mole(TiO$_2$)	300 to 1700°C.

Nonmetals (continued)

Pt vs Cr_2O_3	Very linear from 400 to $1300°C$.
Mo vs $MoSi_2$	Special applications to $1800°C$.
$MoSi_2$ vs WSi_2	0 to $1700°C$ in oxidizing atmospheres. Sensitive to thermal shock.

B. The Elements and Their Presence in Thermocouple Nominal Compositions

The chemical elements which are present in thermocouple nominal compositions given in the text are listed below. In each case the accompanying page numbers refer to locations of couples containing that particular element. A specific thermocouple can be found by selecting the least referenced element given in the required composition and then searching the pages listed. In cases where large numbers of pages are referenced, a process of elimination can be used which consists of rejecting page numbers that are not common to all elements of the thermocouple composition.

Element	Page Numbers	Element	Page Numbers
Ag	66, 76, 88, 89, 91-93, 98, 113, 114	Fe	80, 81, 92, 96, 98-100, 113, 119, 121, 125, 126, 131, 132, 135, 139, 141-144, 158-164, 175
Al	88, 90, 121, 131, 137, 139, 143, 144, 159-162, 163, 164, 175		
		Ge	138, 139
Au	20, 21, 54, 56-59, 61, 63, 68, 77, 81, 90-93, 95, 96, 98-101, 113, 114, 208	Hf	218
		In	80
		Ir	22-24, 54, 57, 58, 64, 68-70, 72, 74, 77, 101-103, 105, 114, 211, 220
B	213, 217-219		
Bi	88, 90, 131, 135, 137, 138, 161	Mg	157, 161
C	131, 159, 160, 162, 163, 208, 211-220	Mn	113, 125, 126, 131, 135, 137, 143, 144, 159, 160, 162, 163
Cb[1]	---	Mo	81, 82, 84, 101, 105, 110, 111, 114, 125, 135, 140, 142, 143, 161-163, 165, 171, 175, 199-201, 203-206, 215, 216, 222
Cd	137, 138		
Cr	98-101, 105, 125, 126, 132, 133, 135, 141, 142, 144, 157-160, 162, 163, 214, 221		
		Nb	98, 158, 161, 175, 201, 203, 216, 218
Co	91, 95, 98, 113, 126, 132, 143, 144, 161, 163, 164	Ni	88, 90, 93, 98-101, 105, 113, 114, 119, 121, 125, 126, 130-133, 135, 137, 139-144, 157-164, 212
Cu	88, 89, 93, 95, 96, 98, 113, 114, 119-135, 161-163, 211, 222		

[1] See Nb.

C. Thermocouple Extension Wires

Thermocouple compositions given in the text which have been used as extension wires are listed below. The page number locations and the thermocouples with which they have been used are also shown.

Extensions	Page Numbers	Used With:
Cu vs Cu-1.78Ni	121	W-5Re vs W-20Re
Cu vs Cu-3Ni	121	Pt-10Rh vs Pt, Mo vs W-26Re, Mo vs Nb
Cu vs Cu-(3.5-8)Sn-(0.05-0.5)P	125	W vs C
Cu vs Fe-(18-20)Cr-(8-12)Ni-(0-2)Mn	125	Ir-40Rh vs Ir, Ir-50Rh vs Ir, Rh-40Ir vs Ir
Cu vs Fe-(17-19)Cr-(9-13)Ni-(0-2)Mn	126	Ir-40Rh vs Ir
Fe vs Cu	119	W-10Re vs W-20Re
Fe vs (82-86)Cu-(4-15)Mn-(2-12)Ni-Fe	126	W-5Re vs W-20Re
Cu-3Ni vs Ni-3Mn-2A1-1Si	131	W-5Re vs W-26Re
Cu-12.3Ni vs Cu-20Ni	131	Pt-10Rh vs Pt
Cu-1.25Ni-0.74Mn-0.01C vs Ni-45Cu	131	(55-83)Pd-(14-34)Pt-(3-15)Au vs Au-(35-40)Pd
Ni-(15-25)Cr vs (96-99)Cu-(0.5-3.0) Ni-(0-1.5)Mn	135	W-3Re vs W-25Re
Ni-10Cr vs Fe	141	W-3Re vs W-25Re
Ni-16Cr-24Fe vs Ni-4.3Si	142	Pt-15Ir vs Pd
Co-23.5Fe vs Co-17Fe	144	W-5Re vs W-26Re
(95-97.5)Ni-(1.5-2.2)Cr-(0-1.2)W-(0.7-1.1) Al-(0.3-1.0)Si vs Ni-(3-5)W	144	W-3Re vs W-25Re
Ni-22Fe-15Cr vs Ni-4.3Si	158	Pt-15Ir vs Pd
Ni-19.7Cr-1.0Si-0.8Ti-0.5Mn-0.3Fe-0.01C vs Ni-3Mn-2Al-1Si	159	W-2(ThO$_2$) vs SiC
Fe-(17-19)Cr-(8-10)Ni-(0-2)Mn-(0-1)Si vs Ni-3Mn-2Al-1Si	160	W-3Re vs W-25Re
Fe-(19-21)Cr-(10-12)Ni-(0-2)Mn-(0-1)Si vs Ni-3Mn-2Al-1Si	160	W-3Re vs W-25Re
Fe-(33.5-34.5)Ni-(19.5-20.5)Cr-(0-0.15)C vs Ni-(4.25-4.75)W	160	Pt-40Rh vs Pt-3Rh

D. Bibliography of Review and General Articles

Authors' last names and publication dates are given. The corresponding article titles and sources can be found using the Reference Section. Reviews, progress reports, tutorial articles, and specifications are included.

Adams and Simpson (1971)
Aleksakhin, Lepin, Lapp, and Bragin (1967
American Society for Testing and Materials (1970)
American Standards Association (1964)
Anderson and Bliss (1971)
Baas and Mai (1971)
Bayley and Turner (1966)
Berman (1971)
Billing (1963)
Borelius (1953)
Burns (1971)
Burton and Weeks (1953)
Caldwell (1962)
Corruccini (1963)
Dike (1954)
Finch (1962)
Freeze (1963)
Gray and Finch (1971)
Howard (1971)
Hust, et al. (1971)

Kostkowski (1960)
Kostkowski and Burns (1963)
Liddiard and Heath (1962)
Miller (1960)
Moffatt (1962)
Rhys (1969)
Rolin and Poupa (1967)
Sagoschen (1961)
Samsonov and Kislyi (1967)
Sanders (1958)
Schulze (1939)
Slaughter and Margrave (1962, 1967)
Sparks and Powell (1971; 1971, p. 19)
Stadnyk and Samsonov (1964)
Starr and Wang (1971)
Stroud (1971)
Toenshoff and Zysk (1971)
Wasan and Gupta (1967)
Wisely (1955)
Zysk (1962, 1964)
Zysk and Robertson (1971)

E. List of Abbreviations
Used in the Text

Å	Ångstrom unit
B	bar
°C	degree Celsius
c	centi-, 10^{-2}, used as in cm = centimeter
eV	electron volt
G	gauss
K	kelvin
k	kilo-, 10^3, used as in kB = kilobar
m	milli-, 10^{-3}, used as in mV = millivolt; meter
mm	millimeter
n	number
n/cm²	number per square centimeter; neutrons per square centimeter (nvt)
n/cm² sec	neutron flux
ppm	parts per million
S	thermoelectric power in units of $\mu V/°C$ or $\mu V/K$
V	volt
μ	micro-, 10^{-6}, used as in μV = microvolt

238

References

The references cited in the text, the tables, and the figures, are given in this section for all chapters. Classification here is by author and date, with the year of publication shown following the author's name(s). Other pertinent information follows in a fixed format which can be understood by studying some of the citations.

In a few instances an author has published more than once during a given year. In this case the text cites the reference by name, year, and some other identifying item. The month or publication page is usually used. For example, the reference shown in the text as Roeser and Wensel (1941, p. 284) refers to the first of the two citations in this section with the heading Roeser, W. F., and H. T. Wensel (1941), where the page number appears last.

Anon. (1937), "The Fitterer Pyrometer," *Iron Age* **140**, July, p. 38.

Anon. (1956), "A Brief Summary of Research on Radiation Effects in Solids," *TID-7517, Pt. 2*, p. 5.

Anon. (1963), "Advance Technical Information, Boron-Graphite/Graphite Thermocouple," *Astro Industries Inc., Bulletin BGT-1*.

Anon. (1965), "The Physical Properties of the Noble Metals," *Engelhard Industries Tech. Bull.* **6**, p. 61.

Anon. (1967), "The Design, Development, Testing and Fabrication of a Thermoelectric Cryogenic Sensor," *NASA-CR-84337*.

Aarset, B., and G. Kjaerheim (1971), "In-Core Applications and Experience with Thermocouples in the Halden Boiling Water Reactor," *Fifth Symp. on Temperature, Washington D.C., Paper T-17*.

Abowitz, G., M. Levy, A. Mountvala, E. Lancaster, V. Klints, and J. Varon (1963), "Thin Film Techniques in Thermoelectricity," *Materials Research Corp., MRC-435*.

Accinno, D. J., and J. F. Schneider (1960), "Platinel, A Noble Metal Thermocouple to Replace Chromel-Alumel," *Engelhard Industries Tech. Bull.* **1**, p. 53.

Accinno, D. J., and J. F. Schneider (1962), "Platinel, A Noble Metal Thermocouple," *Temperature, its Measurement and Control in Science and Industry* vol. 3, pt. 2, Reinhold, New York, p. 195.

Adabaschjan, A. K. (1959), "Temperaturmessung flüssigen Stahles mit Hilfe von Thermoelementen aus Wolfram-Molybdän, Wolfram-Molybdän-Aluminium und Wolfram-Rhenium," *Neue Hütte* 4, p. 252.

Adamenko, A. A. (1969), "Stability of the Thermoelectric Force of Certain Thermocouples in Plastic Deformation," *Ind. Lab.* **35**, p. 1669.

Adams, L. H. (1914), "Calibration Tables for Copper-Constantan and Platinum/Platinum-Rhodium Thermoelements," *J. Am. Chem. Soc.* **36**, p. 65.

Adams, L. H. (1920), "Tables and Curves for Use in Measuring Temperatures with Thermocouples," *Pyrometry,* AIME, New York, p. 165.

Adams, R. K., and R. L. Simpson (1971), "Review Techniques for Thermocouple Emf-Temperature Characteristics," *Fifth Symp. on Temperature, Washington D.C., Paper C-15.*

Aleksakhin, I. A., I. R. Lepin, G. B. Lapp, and B. K. Bragin (1967), "Problems of Investigating Thermoelectrode Alloys, Stable to Oxidation up to 2000°C," *FTD-HT-23-746-67.*

Allen, J. A. (1954), "Evaporated Metal Films," *Rev. Pure Appl. Chem.* **4**, p. 133.

Alvermann, W., and P. Stottmann (1964), "Temperature Measurements with Thermocouples in Combustion Gases," *German Research Inst. for Aeronautics and Space, German Aeronautics and Space Administration Report 64-18.* Also as *NASA-TT-F-9537,* 1965.

American Society for Testing and Materials (1970), "Manual on the Use of Thermocouples in Temperature Measurement," *STP-470.*

American Standards Association (1964), "Temperature Measurement Thermocouples," *ASA-C96.1-1964.*

Anderson, T. M., and P. Bliss (1971), "Tungsten-Rhenium Thermocouples Summary Report," *Fifth Symp. on Temperature, Washington, D.C., Paper T-5.*

Andrew, A., M. R. Jeppson, and H. P. Yockey (1952), "Effect of Cyclotron Irradiation on Some Thermocouple Materials," *Phys. Rev.* **86**, p. 643.

Asamoto, R. R., and P. E. Novak (1967), "Tungsten-Rhenium Thermocouples for Use at High Temperatures," *Rev. Sci. Instr.* **38**, p. 1047.

Ashworth, T., and H. Steeple (1964), "A Thermoelectrically Activated Instrument for the Determination of Levels of Cryogenic Liquids," *J. Sci. Instr.* **41**, p. 782.

Aston, J. G. (1941), "The Use of Copper-Constantan Thermocouples for Measurement of Low Temperatures, Particularly in Calorimetry," *Temperature, its Measurement and Control in Science and Industry,* vol. 1, Reinhold, New York, p. 219.

Aston, J. G., E. Willihnganz, and G. H. Messerly (1935), "Heat Capacities and Entropies of Organic Compounds, I.," *J. Am. Chem. Soc.* **57**, p. 1642.

Baas, P. B. R., and K. Mai (1971), "Trends of Design in Gas Turbine Temperature Sensing Equipment," *Fifth Symp. on Temperature, Washington, D.C., Paper T-36.*

Babbe, E. L. (1964), "Development Program to Increase Thermocouple Reliability for In-Pile Experiments," *TID-7697,* p. 4.3.1.

Bailey, S. B., R. T. Richard, and E. N. Mitchell (1969), "Evaporated Silver-Aluminum Thermocouples for Low Temperature Measurement," *Rev. Sci. Instr.* **40**, p. 1237.

Barber, C. R. (1943), "The Calibration of the Platinum/13% Rhodium-Platinum Thermocouple over the Liquid Steel Temperature Range," *J. Iron Steel Inst.* **147**, p. 205.

Barber, C. R. (1950), "The Emf-Temperature Calibration of Platinum-10% Rhodium/Platinum and Platinum-13% Rhodium/Platinum Thermocouples over the Temperature Range 0-1760°C," *Proc. Phys. Soc. (London)* **63B**, p. 492.

Barber, C. R. (1969), "The Platinum Metals in the Measurement of Temperature," *Platinum Metals* Rev. **13**, p. 65.

Barber, C. R., and L. H. Pemberton (1955), "Silver-Palladium Thermocouples," *J. Sci. Instr.* **32**, p. 486.

Bartoli, A., and G. Papasogli (1882), "Thermoelectric Materials," *Nuovo Cimento,* **12**, p. 181.

Barus, C. (1892), "Thermoelectrics of Platinum-Iridium and of Platinum-Rhodium," *Phil. Mag.* **34**, p. 376.

Baxter, W. G., D. E. Conner, and W. Rauch (1969), "Life Testing Small Diameter Thermocouples as a Function of Thermal Cycling," *General Electric Co., GEMP-737.*

Bayley, F. J., and A. B. Turner (1966), "Bibliography of Heat Transfer Instrumentation," *Aeronautical Research Council (Great Britain) Reports and Memoranda No. 3512.*

Bedford, R. E. (1964), "Reference Tables for Platinum 20% Rhodium/Platinum 5% Rhodium Thermocouples," *Rev. Sci. Instr.* **35**, p. 1177.

Bedford, R. E. (1965), "Reference Tables for Platinum 40% Rhodium/Platinum 20% Rhodium Thermocouples," *Rev. Sci. Instr.* **36**, p. 1571.

Bedford, R. E. (1969), "New Reference Tables for Platinum 10% Rhodium/Platinum and Platinum 13% Rhodium/Platinum Thermocouples, an Interim Report," *24th Annual ISA Conf., Houston, Texas, Paper 69-628. ISA Trans.* 9, 1970, p. 248.

Bedford, R. E. (1971), "Remarks on the International Practical Temperature Scale of 1968," *Fifth Symp. on Temperature, Washington, D.C., Paper S-13.*

Bedford, R. E., C. K. Ma, C. R. Barber, T. R. Chandler, T. J. Quinn, G. W. Burns, and M. Scroger (1971), "New Reference Tables for Platinum 10% Rhodium/Platinum and Platinum 13% Rhodium/Platinum Thermocouples," *Fifth Symp. on Temperature, Washington D.C., Paper T-50.*

Behrndt, K. H. (1960), "An Evaporator for the Controlled Deposition of Alloy Films over Large Substrate Areas," *Am. Vacuum Soc. Symp. Trans.* p. 137.

Bendersky, D. A. (1953), "A Special Thermocouple for Measuring Transient Temperatures," *Mech. Eng.* **75**, p. 117.

Bennett, H. E. (1958), "The Care of Platinum Thermocouples," *Platinum Metals Rev.* **2**, p. 120. Also in *Glass Ind.* **40**, 1959, p. 190.

Bennett, H. E. (1960), "The Pallador Thermocouple," *Platinum Metals Rev.* **4**, p. 66.

Bennett, H. E. (1961), *Noble Metal Thermocouples,* 3rd ed., Johnson, Matthey, London.

Bennett, H. E. (1961), "The Contamination of Platinum Metal Thermocouples," *Platinum Metals Rev.* **5**, p. 132.

Bennett, J. A. (1959), "Investigation into the Binding Energy of Thin Films," *Proc Intern. Conf. on Structure and Properties of Thin Films,* Wiley, New York, p. 58.

Bennett, R. L., and H. L. Hemphill (1963), "Refractory Metal Thermocouples," *ORNL-3417,* p. 167.

Bennett, R. L., H. L. Hemphill, and W. T. Rainey, Jr. (1966), "Thermal Emf Drift of Refractory Metal Thermocouples in Pure and Slightly Contaminated Helium Atmospheres," *WASH-1067, Paper 7.*

Bennett, R. L., H. L. Hemphill, W. T. Rainey, Jr., and G. W. Keilholtz (1966), "Stability of Thermoelectric Materials in a Helium-Graphite Environment," *WASH-1067, Paper 8.*

Bennett, R. L., W. T. Rainey, Jr., and W. M. McClain (1962), "Drift Studies on Chromel-P/Alumel Thermocouples in Helium Atmospheres, *Temperature, its Measurement and Control in Science and Industry,* vol. 3, pt. 2, Reinhold, New York, p. 289.

Bentley, R. E. (1969), "The Emf Criteria for International Practical Temperature Scale Thermocouples," *Metrologia* **5**, p. 26.

Berman, R. (1971), "Gold-Iron Alloys for Low Temperature Thermocouples," *Fifth Symp. on Temperature, Washington D.C., Paper T-46.*

Berman, R., J. C. F. Brock, and D. J. Huntley (1964), "Properties of Gold + 0.03 per cent (at.) Iron Thermoelements Between 1 and 300 K, and Behavior in a Magnetic Field," *Cryogenics* **4**, p. 233.

Berman, R., J. C. F. Brock, and D. J. Huntley (1965), "Dilute Gold-Iron Thermoelements in Low Temperature Thermocouples," *Advances in Cryogenic Engineering,* vol. 10, Plenum Press, New York, p. 233.

Berman, R., and D. J. Huntley (1963), "Dilute Gold-Iron Alloys as Thermocouple Material for Low Temperature Heat Conductivity Measurements," *Cryogenics* **3**, p. 70.

Berman, R., J. Kopp, G. A. Slack, and C. T. Walker (1968), "Magnetic Field Dependence of the Thermoelectric Power for Au + 0.03% Fe at Low Temperatures," *Phys. Letters* **27A**, p. 464.

Berman, R., and S. J. Rogers (1964), "The Thermal Conductivity of Solid He^4 in the γ Phase," *Phys. Letters* **9**, p. 115.

Bertodo, R. J. (1963), "A Thermocouple for the Measurement of Gas Temperatures up to 2000°C," *Proc. Inst. Mech. Engrs. (London)* **177**, p. 603.

Bianchi, G., P. Lupoli, and S. Moretti (1966), "Behavior of Irradiated Thermocouples, Experimental Investigations," *Comitato Nazionale Energia Nucleare, Rome, RT/ING (66)* **14.**

Bidwell, C. C. (1914), "Note on a Thermojunction of Carbon and Graphite," *Phys. Rev.* **3**, p. 450.

Billing, B. F. (1963), "Thermocouples, a Critical Survey," *Royal Aircraft Establishment, Farnborough, England, RAE-TN-CPM-18.*

Black, F. S., and H. W. Wilken (1963), "High Temperature Sensors for a Re-Entry Research Vehicle," *Soc. Automotive Engrs., National Aeronautic and Space Engineering and Manufacturing Meeting, Los Angeles, Paper 750K.*

Blackburn, G. F., and F. R. Caldwell (1962), "Reference Tables for 40% Iridium-60% Rhodium versus Iridium Thermocouples," *Temperature, its Measurement and Control in Science and Industry,* vol. 3, pt. 2, Reinhold, New York, p. 161. Also in *J. Res. Natl. Bur. Std. (U.S.)* **66C**, 1962, p. 1.

Blackburn, G. F., and F. R. Caldwell (1964), "Reference Tables for Thermocouples of Iridium-Rhodium Alloys versus Iridium," *J. Res. Natl. Bur. Std. (U.S.)* **68C**, p. 41.

Bliss, P. (1965), "Status of Instrument Development for the Snap-50 Reactor Test Program," *Pratt and Whitney Aircraft, PWAC-473.*

Bliss, P. (1971), "Fabrication of High Reliability Sheathed Thermocouples," *Fifth Symp. on Temperature, Washington, D.C., Paper T-25.*

Bliss, P., and S. Fanciullo (1965), "High Temperature Thermometry at Pratt and Whitney Aircraft-CANEL," *PWAC-462.*

Bliss, P., and S. Fanciullo (1966), "High Temperature Thermometry at CANEL," *IEEE Nucl. Sci. Trans.* **NS-13**, p. 643.

Bloch, D., and F. Chaissé (1967), "Effect of Pressure on Emf of Thermocouples," *J. Appl. Phys.* **38**, p. 409.

Bondareva, A. P. (1969), "Method for Calibrating Copper-Constantan Thermocouples at Negative Temperatures by Means of the Boiling Point of Water," *Meas. Tech.* (10), p. 1428.

Boorman, C. (1950), "The Effect of Neutron Flux on the Thermoelectric Power of Thermocouples and on the Resistivity of Platinum and Nickel," *United Kingdom Atomic Energy Authority, AERE-E/R-572.*

Borelius, G. (1953), "Thermoelectricity at Low Temperatures," *Physica* 19, p. 807.

Borelius, G., W. H. Keesom, C. H. Johansson, and J. O. Linde (1932), "Thermoelectric Forces Down to Temperatures Obtainable with Liquid or Solid Hydrogen and with Liquid Helium," *Proc. Acad. Sci., Amsterdam* 35, p. 15. Also as *Leiden Comm. No. 217d.*

Bostwick, W. E. (1962), "A Note on Platinum/Platinum-Rhodium Thermocouple Uncertainty," *Trans. Inst. Radio Engrs.* NS-9, p. 253.

Bragin, B. K., and R. N. Tetyueva (1964), "Thermoelectric Uniformity of Chromel, Copel, Alumel, and Copper Wire at Low Temperatures," *Meas. Tech.* (6), p. 498.

Brenner, B. (1941), "Changes in Platinum Thermocouples due to Oxidation," *Temperature, its Measurement and Control in Science and Industry,* vol. 1, Reinhold, New York, p. 1281.

Bridgeman, P. W. (1918), "Thermoelectric Force, Peltier Heat, and Thomson Heat under Pressure," *Proc. Am. Acad. Arts Sci.* 53, p. 269.

Briggs, N. H. (1971), "Metal Sheathed, Compacted Ceramic Oxide Insulated Thermocouples —General Discussion with Emphasis on Compatibility Factors," *Fifth Symp. on Temperature, Washington, D.C.,* Paper T-23.

Briggs, N. H., and W. W. Johnston, Jr. (1966), "Thermal Emf Drift of High-Temperature Thermocouples in Helium and Argon," *WASH-1067, Paper 6.*

Bristol, W. H. (1908), "Thermocouples of Graphite," *Engineering* 86, p. 37. "Thermoelectric Couple," *U.S. Patent 885,430,* (1908).

Brown, D. A. H., R. P. Chasmar, and P. B. Fellgett (1953), "The Construction of Radiation Thermocouples using Semiconducting Thermoelectric Materials," *J. Sci. Instr.* 30, p. 195.

Browning, W. E., Jr., and C. E. Miller, Jr. (1962) "Calculated Radiation Induced Changes in Thermocouple Composition," *Temperature, its Measurement and Control in Science and Industry,"* vol. 3, pt. 2, Reinhold, New York, p. 271.

Buchanan, J. (1885), "On the Thermoelectric Position of Carbon," *Phil. Mag.* 20, p. 117.

Bullis, L. H. (1963), "Vacuum Deposited Thin Film Thermocouples for Accurate Measurement of Substrate Surface Temperature," *J. Sci. Instr.* 40, p. 592.

Bundy, F. P. (1961), "Effect of Pressure on Emf of Thermocouples," *J. Appl. Phys.* 32, p. 483.

Burley, N. A. (1969), "Solute Depletion and Thermo-Emf Drift in Nickel-Base Thermocouple Alloys," *J. Inst. Metals* 97, p. 252.

Burley, N. A. (1970), "Cyclic Thermo-Emf Drift in Nickel-Chromium Thermocouple Alloys Attributable to Short Range Order," *Australian Defense Scientific Service, DSL-353.*

Burley, N. A. (1971), "Nicrosil and Nisil: Highly Stable Nickel-Base Alloys for Thermocouples," *Fifth Symp. on Temperature, Washington, D.C.,* Paper T-4.

Burns, G. W. (1971), "Pt-67 — A New Platinum Thermoelectric Reference Standard," *Fifth Symp. on Temperature, Washington D.C.,* Paper T-51.

Burns, G. W., and J. S. Gallagher (1966), "Reference Tables for the Pt-30 Per Cent Rh versus Pt-6 Per Cent Rh Thermocouple," *J. Res. Natl. Bur. Std. (U.S.)* 70C, p. 89.

Burns, G. W., and W. S. Hurst (1970), "An Investigation of W-3%Re and W-25%Re Thermoelements in Vacuum, Argon, and Hydrogen," *NASA-CR-72639.*

Burns, G. W., and W. S. Hurst (1971), "Studies of the Performance of W-Re Type Thermocouples," *Fifth Symp. on Temperature, Washington D.C.,* Paper T-7.

Burton, E. J., and D. J. Weeks (1953), "Progress Review No. 27: Temperature Measurement," *J. Inst. Fuel* 26, p. 260.

Caldwell, F. R. (1962), "Thermocouple Materials," *Natl. Bur. Std. Monograph 40.* Also in *Temperature, its Measurement and Control in Science and Industry,* vol. 3, pt. 2, Reinhold, New York, 1962, p. 81.

Calkins, G..D., and P. Schall (1960), "Radiation Damage, Miscellaneous Materials," *Reactor Handbook,* 2nd ed., vol. 1, Materials, C. R. Tipton, ed., Interscience, New York, p. 79.

Callcut, V. A. (1965), "Aging of Chromel-Alumel Thermocouples," *United Kingdom Atomic Energy Authority, TRG-1021 (R/X).*

Campari, M., and S. Garribba (1971), "The Behavior of Type K Thermocouples in Temperature Measurement: The Chromel-P/Alumel Thermocouples," *Rev. Sci. Instr.* **42**, p. 644.

Carpenter, F. D., N. L. Sandefur, R. J. Grenda, and J. S. Steibel (1971), "Emf Stability of Chromel-Alumel and Tungsten-3% Rhenium/Tungsten-25% Rhenium Sheathed Thermocouples in a Neutron Environment," *Fifth Symp. on Temperature, Washington D.C., Paper T-14.*

Carr, K. R. (1971), "Testing of Ceramic Insulator Compaction and Thermoelectric Inhomogeneity in Sheathed Thermocouples," *Fifth Symp. on Temperature, Washington D.C., Paper T-26.*

Carroll, R. M., and P. E. Reagan (1966), "In-Pile Performance of High Temperature Thermocouples," *WASH-1067, Paper 2.*

Carter, F. E. (1950), Discussion of Troy and Steven (1950), *Trans. Am. Soc. Metals* **42**, p. 1151.

Cartwright, C. H. (1932), "Tellurium-Bismuth Radiation Thermocouple," *Rev. Sci. Instr.* **3**, p. 73.

Cartwright, C. H. (1932), "Construction of Thermo-Relay Amplifiers," *Rev. Sci. Instr.* **3**, p. 221.

Cartwright, C. H. (1933), "Radiation Thermopiles for Use at Liquid Air Temperatures," *Rev. Sci. Instr.* **4**, p. 382.

Chand, R., and H. F. Rosson (1965), "Local Heat Flux to a Water Film Flowing Down a Vertical Surface," *I EC Fundamentals* **4**, p. 356.

Chaston, J. C. (1957), "A Thermocouple for High Temperatures," *Platinum Metals Rev.* **1**, p. 20.

Chaston, J. C., R. A. Edwards, and F. M. Lever (1947), "Embrittlement of Platinum/Platinum-Rhodium Thermocouples," *J. Iron Steel Inst.* **155**, p. 229.

Chaussain, M. (1952), "Étude sur les couples thermo-électriques platine-platine rhodié et applications industrielles," *Fonderie* (77), p. 2955.

Chopra, K. L., S. K. Bahl, and M. R. Randlett (1968), "Thermopower in Thin Film Copper-Constantan Couples," *J. Appl. Phys.* **39**, p. 1525.

Christensen, J. A., and H. M. Ferrari (1966), "In-Pile Refractory Thermocouple Experience at Westinghouse Atomic Power," *WASH-1067, Paper 30.*

Clark, H. T. (1946), "Recent Developments in the Pyrometry of Liquid Iron and Steel, *Iron and Steel Engineer* **23**, Feb., p. 55.

Clark, R. B. (1962), "Time-Temperature Effect on Jet Engine Thermocouple Accuracy and Reliability," *Soc. Automotive Engrs., National Aeronautic Meeting, New York, Paper 524K.*

Clark, R. B. (1964), "Calibration and Stability of W/Re Thermocouples to 2760°C (5000°F)," *19th Annual ISA Conf. Proc., Paper 16.10-4-64.* **19**, pt. 2.

Clark, R. B., and W. C. Hagel (1960), "High Output Noble Metal Thermocouples and Matching Lead Wire," *TID-7586,* pt. 1, p. 37.

Cochrane, J. (1970, 1971), "Relationship of Chemical Composition to the Electrical Properties of Platinum," *Engelhard Industries Tech. Bull.* **11**, p. 58, 1970. *Fifth Symp. on Temperature, Washington D.C., 1971, Paper T-2.*

Cook, G. E., and W. M. McCampbell (1968), "Intrinsic Thermocouple Monitors Welding," *Metal Progress* **93**, June, p. 176.

Cormier, M., and F. Claisse (1967), "Temperature Measurement During Quenching of Titanium," *Can. Met. Quart.* **6**, p. 369.

Corruccini, R. J. (1951), "Annealing of Platinum for Thermometry," *J. Res. Natl. Bur. Std. (U.S.)* **47**, p. 94.

Corruccini, R. J. (1963), "Temperature Measurements in Cryogenic Engineering," *Advances in Cryogenic Engineering,* vol. 8, Plenum Press, New York, p. 315.

Corruccini, R. J., and H. Shenker (1953), "Modified 1913 Reference Tables for Iron-Constantan Thermocouples," *J. Res. Natl. Bur. Std. (U.S.)* **50**, p. 229.

Cox, J. E. (1963), "High Vacuum Thermocouple," *Rev. Sci. Instr.* **34**, p. 931.

Crisp, R. S., and W. G. Henry (1964), "The Design of a Low Temperature Thermocouple Material," *Cryogenics* **4**, p. 361.

Cunningham, G. W., and W. H. Goldthwaite (1966), "Contribution to the Meeting on High Temperature Thermometry," *WASH-1067, Paper 25.*

Dahl, A. I. (1941), "The Stability of Base Metal Thermocouples in Air from 800 to 2200°F," *Temperature, its Measurement and Control in Science and Industry,* vol. 1, Reinhold, New York, p. 1238.

Dalle Donne, M. (1964), "Tests and Data Concerning Platinel, a New High Temperature Thermocouple," *Engelhard Industries Tech. Bull.* **5**, p. 5.

Dallman, A. C. (1964), "Mechanical Reliability and Thermoelectrical Stability of Noble Metal Thermocouples at 2600°F Temperature and Dose Rates up to 10^{20} nvt," *SEG-TDR-64-7.*

Dalton, J. E. (1969), "The Measurement of Temperature of Soaking Pit Waste Gases," *Platinum Metals Rev.* **13**, p. 139.

Danishevskii, S. K. (1959), "Selection and Calibration of Tungsten and Molybdenum Wire for Thermocouples," *Meas. Tech.* (5), p. 333.

Danishevskii, S. K. (1969), "Rhenium-Tungsten Alloys as Material for High-Temperature Thermocouples," *FTD-MT-24-11-69. Trans. All-Union Conf. on the Problem of Rhenium, Moscow, 1961,* p. 162 (translation).

Danishevskii, S. K., S. I. Ipatova, E. I. Pavlova, and N. I. Smirnova (1964), "Thermocouples Made of Tungsten Alloys with Rhenium for Measuring Temperature up to 2500°C," *Ind. Lab.* **29**, p. 1242.

Danishevskii, S. K., S. I. Ipatova, P. P. Oleinikov, L. D. Oleinikova, E. I. Pavlova, N. I. Smirnova, and L. I. Trakhtenberg (1967), "Molybdenum-Rhenium Alloy Thermocouples," *Meas. Tech.* (4), p. 508.

Danishevskii, S. K., L. D. Oleinikova, P. P. Oleinikov, N. I. Smirnova, and L. I. Trakhtenberg (1968), "Calibration Characteristic of Type VR 5/20 Tungsten-Rhenium Thermocouples," *Meas. Tech.* (7), p. 877.

Darling, A. S. (1961), "Rhodium-Platinum Alloys," *Platinum Metals Rev.* **5**, p. 58.

Darling, A. S., and G. L. Selman (1971), "Some Effects of Environment on the Performance of Noble Metal Thermocouples," *Fifth Symp. on Temperature, Washington D.C., Paper T-29.*

Darling, A. S., G. L. Selman, and R. Rushforth (1970, 1971)," *Platinum and the Refractory Oxides,"* pts. I–IV, *Platinum Metals Rev.* **14**, pp. 54, 95, 124; **15**, p. 13.

Dau, G. J., R. R. Bourassa, and S. C. Keeton (1968), "Nuclear Radiation Dose Rate Influence on Thermocouple Calibration," *Nucl. Applic.* **5**, p. 322.

Dauphinee, T. M., D. K. C. MacDonald, and W. B. Pearson (1953), "The Use of Thermocouples for Measuring Temperatures below 70 K, A New Type of Low Temperature Thermocouple," *J. Sci. Instr.* **30**, p. 399.

Davies, D. A. (1960), "Two Thermocouples Suitable for Measurement of Temperatures up to 2800°C," *J. Sci. Instr.* **37**, p. 15.

Davies, J. M. (1970), "Skin Simulants for Studies of Protection Against Intense Thermal Radiation," *Rev. Sci. Instr.* **41**, p. 1040.

Day, G. W., O. L. Gaddy, and R. J. Iversen (1968), "Detection of Fast Infrared Laser Pulses with Thin Film Thermocouples," *Appl. Phys. Letters* **13**, p. 289.

Dike, P. H. (1954), *Thermoelectric Thermometry,* Leeds and Northrup Co. Philadelphia, Pa.

Dobovišek, B., and A. Rosina (1968), "Analysis of the Influence of Various Atmospheres on the Thermoelectric Power of the NiCr-Ni Thermocouple," *Mining and Metallurgical Quarterly* (3S), p. 5.

Driesner, A. R., C. P. Kempter, C. E. Landahl, C. A. Linder, and T. E. Springer (1962), "High Temperature Thermocouples in the Rover Program," *Trans. Inst. Radio Engrs.* **NS-9** p. 247.

Driesner, A. R., C. P. Kempter, C. E. Landahl, C. A. Linder, and T. E. Springer (1962), "High Temperature W/W-25 Re Thermocouples," *Instr. Control Systems* **35**, May, p. 105.

Droege, J. W., N. E. Miller, M. E. Schimek, V. E. Wood, and J. J. Ward (1968), "Refractory Metal Thermocouples in Nuclear and High Temperature Applications," *Battelle Memorial Inst., BMI-X-10246.*

Droege, J. W., M. M. Schimek, and J. J. Ward (1971), "Chemical Compatibility in the Use of Refractory Metal Thermocouples," *Fifth Symp. on Temperature, Washington D.C., Paper T-32.*

Droms, C. R., and A. I. Dahl (1955), "Iridium versus Iridium-Rhodium Thermocouples for Gas Temperature Measurements up to 3500°F," *ASME, Proc. Joint Conf. on Combustion, Boston, London,* p. 330.

Druzhinina, I. P., T. M. Vladimirskaya, and A. A. Fraktovnikova (1966), "Thermoelectric Properties of Vanadium," *Meas. Tech.* (8), p. 1032.

Dutta, A. K., and R. Bhattacharya (1965), "Copper-Molybdenite Thermocouple," *Indian J. Phys.* **39**, p. 251.

Dworschak, F., J. Neuhäuser, H. Schuster, H. Wollenberger, and J. Wurm (1970), "A Specimen Holder for Low Temperature Electron Irradiation of Metallic Resistivity Specimens," *Rev. Sci. Instr.* **41**, p. 64.

Efremova, R. I., N. V. Kuzkova, L. N. Levina, and É. V. Matizen (1963), "Temperature Measurements by Means of Copper-Constantan Thermocouples," *Meas. Tech.* (3), p. 214.

Ehringer, H. (1954), "Über die lebensdauer von Pt-Rh Thermoelementen," *Metall.* **8**, p. 596.

Ehringer, H., C. Mongini-Tamagnini, and C. Ponti (1966), "Thermocouple Composition Changes Due to Neutron Irradiation," *European Atomic Energy Community, EUR-3156e.*

Ergardt, N. N. (1968), "Temperature Measurements with Tungsten-Rhenium and Tungsten-Molybdenum Thermocouples and their Calibration Methods," *Meas. Tech.* (10), p. 1335.

Eshaya, A. M. (1963), "High Temperature Thermocouple," *U.S. Patent 3,077,505.*

Evans, R. A., and J. Holt (1967), "Calibration Drift Effects in Nickel-Alloy Thermocouples," *United Kingdom Atomic Energy Authority, TRG-1510(R)*.

Fairchild, C. O., and H. M. Schmitt (1922), "Life tests of Platinum/Platinum-Rhodium Thermocouples," *Chem. Met. Eng.* **26**, p. 158.

Fanciullo, S. (1964), "Thermocouple Development, Lithium Cooled Reactor Experiment," *Pratt and Whitney Aircraft, PWAC-422*.

Fanciullo, S. (1965), "Drift and Endurance Testing of Chromel-Alumel, W-5Re/W-26Re and Mo/W-26Re Thermocouples at 1950-2000°F for 10,000 Hours," *Pratt and Whitney Aircraft, PWAC-454*.

Farrow, R. L., and A. P. Levitt (1964), "Tungsten/Tungsten-Rhenium Thermocouples in a Carbon Atmosphere," *AMRA-TR-64-12*.

Fedorovskii, M. L. (1969), "Measuring Temperature of Liquid Steel with Immersion Thermocouples," *Meas. Tech.* (5), p. 756.

Feigenbutz, L. V. (1968), "Thermoelectric Potentials for a Tungsten-Tantalum Thermocouple from 0 to 420°C," *J. Phys. E* **1**, p. 489.

Fenton, A. W. (1967), "Metal-Sheathed Mineral-Insulated Base-Metal Thermocouples," *United Kingdom Atomic Energy Authority, TRG-1404(R), Paper 5*.

Fenton, A. W. (1969), "Errors in Thermoelectric Thermometers," *Proc. Inst. Elect. Engrs. (London)* **116**, p. 1277.

Fenton, A. W. (1971), "The Travelling Gradient Approach to Thermocouple Research," *Fifth Symp. on Temperature, Washington D.C., Paper T-43*.

Ferry, P. B. (1964), Addendum to "Development Program to Increase Thermocouple Reliability for In-Pile Experiments," *TID-7697*, p. 4.3.6.

Feussner, O. (1927), "Edelmetallthermoelemente mit hoher Thermokraft," *Elektrotech. Z.* **48**, p. 535.

Feussner, O. (1933), "New Noble Metal Thermoelements for Very High Temperatures," *Elektrotech. Z.* **54**, p. 155.

Finch, D. I. (1962), "General Principles of Thermoelectric Thermometry," *Temperature, its Measurement and Control in Science and Industry*, vol. 3, pt. 2, Reinhold, New York, p. 3.

Fink, C. G. (1912), *Trans. Am. Electrochem. Soc.* **22**, p. 501.

Fink, C. G. (1913), "Applications of Ductile Tungsten," *J. Ind. Eng. Chem.* **5**, p. 8.

Finnemore, D. K., J. E. Ostenson, and T. F. Stromberg (1965), "Secondary Thermometer for the 4 to 20 K Range," *Rev. Sci. Instr.* **36**, p. 1369.

Fischer, W. A., and G. Lorenz (1958), "Untersuchungen zur Entwicklung eines oxydishen Thermoelementes," *Archiv für das Eisenhüttenwesen* **29**, p. 293.

Fitterer, G. R. (1933), "A New Thermocouple for the Determination of Temperatures up to at Least 1800°C," *Trans. AIME* **105**, p. 290.

Fitterer, G. R. (1936), "Some Metallurgical Applications of the C-SiC Thermocouple," *Trans. AIME* **120**, p. 189.

Fitterer, G. R. (1941), "Pyrometry of Liquid Steels and Pig Irons," *Temperature, its Measurement and Control in Science and Industry*, vol. 1, Reinhold, New York, p. 946.

Fitts, R. B., J. L. Miller, Jr., and E. L. Long, Jr. (1971), "Observations on Tungsten-Rhenium Thermocouples used In-Reactor in (U, Pu)O_2 Fuel Pins," *Fifth Symp. on Temperature, Washington, D.C., Paper T-21*.

Fleischner, P. L. (1971), "The Utilization of Beryllia, Hafnia, and Thoria Ceramics as High Temperature Thermocouple Insulation," *Fifth Symp. on Temperature, Washington, D.C., Paper T-9.*

Flemons, R. S., and A. D. Lane (1971), "Thermocouples for Surface Temperature Measurement in Multielement Nuclear Fuel Bundles," *Fifth Symp. on Temperature, Washington D.C., Paper T-19.*

Ford, W. D. (1951), "Method of Coating Thermocouples," *U.S. Patent 2,571,700.*

Franks, E. (1962), "High Temperature Thermocouples Using Nonmetallic Members," *Temperature, its Measurement and Control in Science and Industry,* vol. 3, pt. 2, Reinhold, New York, p. 189.

Freeman, R. H., and J. Bass (1970), "An ac System for Measuring Thermopower," *Rev. Sci. Instr.* **41**, p. 1171.

Freeman, R. J. (1962), "Thermoelectric Stability Platinum vs Platinum-Rhodium Thermocouples," *Temperature, its Measurement and Control in Science and Industry,* vol. 3, pt. 2, Reinhold, New York, p. 201.

Freeze, P. D. (1963), "Review of Recent Development of High Temperature Thermocouples," *ASME, Winter Annual Meeting, Philadelphia, Paper 63-WA-212.*

Freeze, P. D., F. R. Caldwell, and E. R. Davis (1962), "Reference Tables for the Palladium vs Platinum-15% Iridium Thermocouple," *ASD-TDR-62-525.*

Freeze, P. D., and L. O. Olsen (1962), "Thermoelectric and Mechanical Stability of Platinel II Thermocouples in Oxidizing Atmospheres," *ASD-TDR-62-835.*

Fries, R. J., J. E. Cummings, C. G. Hoffman, and S. A. Daily (1969), "Compatibility of Some High Temperature Thermocouple Materials in a Carbon-Hydrogen Environment," *J. Nucl. Mat.* **32**, p. 171.

Funston, E. S., and W. C. Kuhlman (1966), "High Temperature Thermocouple and Electrical Materials Research," *High Temperature Materials Program Progress Report No. 63, General Electric Co. GEMP-63.*

Funston, E. S., and W. C. Kuhlman (1967), "High Temperature Thermocouple and Electrical Materials Research," *AEC Fuels and Materials Development Program Progress Report No. 69, General Electric Co., GEMP-69,* p. 63.

Fuschillo, N. (1954), "Inhomogeneity Emf's in Thermoelectric Thermometers," *J. Sci. Instr.* **31**, p. 133.

Fuschillo, N. (1957), "A Low Temperature Scale from 4 to 300 K in Terms of a Gold-Cobalt versus Copper Thermocouple," *J. Phys. Chem.* **61**, p. 644.

Gatward, W. A. (1941), Discussion of Dahl (1941, p. 1238) in *Temperature, its Measurement and Control in Science and Industry,* vol. 1, Reinhold, New York, p. 1260.

Gaylord, A. M., and W. A. Compton (1968), "High Temperature Sensors for Small Gas Turbines," *USA-AVLABS-TR-67-76.*

Gee, L. J. (1963), "Thermocouple Probe," *U.S. Patent 3,116,168.*

Geibel, W. (1910), "Uber einige elektrische und mechanische Eigenshaften von Edelmetall-Legierungen," *Z. Anorg. Chem.* **69**, p. 38.

General Electric Co. (Undated), "Tables of Thermocouple Characteristics," *Instrument Dept., Publication 198-4559K99-001* (formerly as *GEI-44592*).

Gerhardt, R. C. (1962), *Vacuum Deposited Thermocouples,* M. S. Thesis, Univ. Denver.

Getting, I. C., and G. C. Kennedy (1970), "Effect of Pressure on the Emf of Chromel-Alumel and Platinum-Platinum 10% Rhodium Thermocouples," *J. Appl. Phys.* **41**, p. 4552.

Gier, J. T., and L. M. K. Boelter (1941), "The Silver-Constantan Plated Thermopile," *Temperature, its Measurement and Control in Science and Industry,* vol. 1, Reinhold, New York, p. 1284.

Glawe, G. E. (1970), "Thermal Electromotive Force Change for Some Noble and Refractory Metal Thermocouples at 1600 K in Vacuum, Air, and Argon," *NASA-TN-D-7027.*

Glawe, G. E., and C. E. Shepard (1954), "Some Effects of Exposure to Exhaust Gas Streams on Emittance and Thermoelectric Power of Bare Wire Platinum-Rhodium/Platinum Thermocouples," *NACA-TN-3253.*

Glawe, G. E., and A. J. Szaniszlo (1971), "Long Term Drift of Some Noble and Refractory Metal Thermocouples at 1600 K in Air, Argon, and Vacuum," *Fifth Symp. on Temperature, Washington, D.C., Paper T-31.*

Goedecke, W. (1931), *Festschrift der Platinschmelze,* G. Siebert, Hanau a.M., p. 72.

Goldschmidt, H. J., and T. Land (1947), "An X-Ray Investigation of the Embrittlement of Platinum and Platinum-Rhodium Wires," *J. Iron Steel Inst.* **155**, p. 221.

Goodier, B. G., R. J. Fries, and C. R. Tallman (1971), "High Temperature Core Thermocouple Development for the Nuclear Rocket Engine Program (Rover)," *Fifth Symp. on Temperature, Washington D.C., Paper T-20.*

Gottlieb, A. J. (1971), "Corrections for Base Metal Thermocouples and Thermoelements due to the Adoption of IPTS-68 and Platinum 67," *Fifth Symp. on Temperature, Washington, D.C., Paper T-49.*

Gottschlich, R., and M. Bellezza (1966), "Emf Deviation of Chromel-Alumel Thermocouples Caused by Neutron Irradiation," *European Atomic Energy Community, EUR-3257.e.*

Grant, N. J., and D. S. Bloom (1950), Discussion of Troy and Steven (1950), *Trans. Am. Soc. Metals* **42**, p. 1152.

Gray, W. T., and D. I. Finch (1971), "State-of-the-Art Accuracy of Temperature Measurement," *Fifth Symp. on Temperature, Washington, D.C., Paper C-13.* "How Accurately can Temperature be Measured?" *Physics Today* **24**, Sept., p. 32.

Green, S. J., and T. W. Hunt (1962), "Accuracy and Response of Thermocouples for Surface and Fluid Temperature Measurements," *Temperature, its Measurement and Control in Science and Industry,* vol. 3, pt. 2, Reinhold, New York, p. 695.

Greenaway, H. T., S. T. M. Johnstone, and M. K. McQuillan (1951), "High Temperature Thermal Analysis using the Tungsten-Molybdenum Thermocouple," *J. Inst. Metals* **80**, p. 109. Also in *Industrial Heating,* **19**, 1952, p. 2270.

Greenberg, H. J. (1971), "Temperature Measurement in the Glass Industry," *Fifth Symp. on Temperature, Washington D.C., Paper T-39.*

Greenberg, H. J., and E. D. Zysk (1963), "Applied Research, Fabrication, and Testing of 2300°F Thermocouple for Air Breathing Propulsion Systems," *ASD-TDR-62-891.*

Gross, J., and C. B. Griffith (1963), "A Thermocouple for Molten Steel," *Metal Progress* **83**, June, p. 106.

Gross, P. M., D. R. Dohr, R. G. Drewes, and R. T. Luedeman (1966), "An Advanced High Temperature Thermocouple for Use on Aerospace Vehicles," *AFFDL-TR-66-24.*

Gross, P. M., J. Gugliotta, K. F. Krysiak, and L. B. Gross (1964), "A Re-Entry Thermocouple for Use up to 4500°F," *ISA Trans.* **3**, p. 305.

Guettel, C. L. (1956), "A New Thermocouple for Service in Reducing Atmospheres," *ASTM Bull.* (216), p. 64. Also in *Metal Progress* **69**, April, p. 89.

Haase, G., and G. Schneider (1956), "Untersuchungen an Thermoelementen aus dem System Iridium-Rhenium," *Z. Physik* **144**, p. 256.

Hall, B. F., Jr., and N. F. Spooner (1963), "Application and Performance Data for Tungsten-Rhenium Alloy Thermocouples," *Soc. Automotive Engrs., National Aeronautic and Space Engineering and Manufacturing Meeting, Los Angeles, Paper 750C.*

Hall, B. F., Jr., and N. F. Spooner (1964), "Temperature Measurement in a Graphite Environment from $1600°C$ to $2500°C$," *19th Annual ISA Conf. Proc.* 19, paper 16.13-3-64.

Hall, B. F., Jr., and N. F. Spooner (1965), "Study of High Temperature Thermocouples," *AFCRL-65-251.*

Hall, J. A. (1966), "Fifty Years of Temperature Measurement," *J. Sci. Instr.* 43, p. 541.

Hancock, N. H. (1967), "The Use of Noble and Refractory Metal Thermocouples in Experimental Irradiation Equipment," *Proc. Internat. Symp. on Developments in Irradiation Capsule Technology,* D. R. Hoffman, ed., Joint AEC/GE Meeting.

Handbook of Chemistry and Physics (1960), 41st ed., Chemical Rubber Publishing Co., Cleveland, Ohio, p. 2640.

Hanneman, R. E., and H. M. Strong (1965), "Pressure Dependence of the Emf of Thermocouples to $1300°C$ and 50 kBar," *J. Appl. Phys.* 36, p. 523.

Hanneman, R. E., and H. M. Strong (1966), "Pressure Dependence of the Emf of Thermocouples," *J. Appl. Phys.* 37, p. 612.

Harris, L. (1934), "Thermocouples for the Measurement of Small Intensities of Radiations," *Phys Rev.* 45, p. 635.

Harris, L. (1946), "Rapid Response Thermopiles," *J. Opt. Soc. Am.* 36, p. 597.

Harris, L., and E. A. Johnson (1934), "The Technique of Sputtering Sensitive Thermocouples," *Rev. Sci. Instr.* 5, p. 153.

Hartwig, F. W. (1957), "Development and Application of a Technique for Steady-State Aerodynamic Heat-Transfer Measurements," *Calif. Inst. Tech., Guggenheim Aero. Lab., Hypersonic Research, Memo. No. 37.*

Hatfield, H. S., and F. J. Wilkins (1950), "A New Heat-Flow Meter," *J. Sci. Instr.* 27, p. 1.

Hatfield, W. H. (1917), "Notes on Pyrometry from the Standpoint of Ferrous Metallurgy," *Trans. Faraday Soc.* 13, p. 289.

Hawkings, R. C. (1964), "Neutron Flux Monitors and Thermocouples for In-Core Reactor Measurements," *Symp. on In-Core Instrumentation, Oslo, Paper CRL-85.* Also as *AECL-2033,* 1964.

Heckelman, J. D., and R. P. Kozar (1971), "Measured Drift of Irradiated and Unirradiated W3%Re/W25%Re Thermocouples at a Nominal 2000 K," *Fifth Symp. on Temperature, Washington D.C., Paper T-13.*

Helm, J. W. (1967), "Irradiation Effects on Thermocouples," *Proc. Internat. Symp. on Developments in Irradiation Capsule Technology,* D. R. Hoffman, ed., Joint AEC/GE Meeting.

Hendricks, J. W., and D. L. McElroy (1966), "High Temperature High Vacuum Thermocouple Drift Tests," *WASH-1067, Paper 9.*

Hendricks, J. W., and D. L. McElroy (1966), "High Temperature, High Vacuum Thermocouple Drift Tests," *Environmental Quarterly* 13, p. 34.

Henning, F. (1938), *Wärmetechnische Richtwerte,* VDI Verlag, Berlin, p. 13.

Hensel, R., and M. Göhler (1964), "Erprobung eines Wolfram-Molybdän Tauchthermoelements zur Temperaturmessung in Stahlschmelzen," *Neue Hütte* 9, p. 237.

Hesse, L. (1969), "Miniature Sheathed Thermocouples in Reactor Engineering," *Kerntechnik,* 11, p. 281.

Hessinger, P. S., K. H. Styhr, R. L. Sharkitt, and W. C. Allen (1965), "Development and Fabrication of Prototype 3000°C Thermocouple Gages," *National Beryllia Corp.* (report).

Hicks, W. T. (1969), "Thermoelectric Alloy of Gold and Nickel," *U.S. Patent 3,438,819.*

Hicks, W. T., and H. Valdsaar (1963), "High Temperature Thermoelectric Generator," *E. I. DuPont deNemours and Co., Quarterly Report,* Oct. 1–Dec. 31, 1963.

Hill, J. S. (1961), "The Use of Fibro Platinum in Thermocouple Elements," *Engelhard Industries Tech. Bull.* **2**, p. 85.

Hill, J. S. (1962), "Fibro Platinum for Thermocouple Elements," *Temperature, its Measurement and Control in Science and Industry,* vol. 3, pt. 2, Reinhold, New York, p. 157.

Hill, J. S. (1963), "Thermocouple," *U.S. Patent 3,099,575.*

Hill, J. S., and H. J. Albert (1963), "Loss of Weight of Platinum, Rhodium, and Palladium at High Temperatures," *Engelhard Industries Tech. Bull.* **4**, p. 59.

Hill, R. A. W. (1957), "Rapid Measurement of Thermal Conductivity by Transient Heating of a Fine Thermojunction," *Proc. Roy. Soc. (London)* **239A**, p. 476.

Hoffmann, F. (1909), "The Measurement of Very High Temperatures," *Ber. Ver. Fabr. Feuerfest. Prod.* **29**, p. 45.

Hoitink, N. C., R. C. Weddle, and D. C. Thompson (1970), "Effects of Fast Neutron Irradiation on Thermocouples," *IEEE Trans. Nucl. Sci.* **NS-17**, p. 262.

Holmes, R. R. (1961), "Development of High Temperature Sensors," *Pratt and Whitney Aircraft, PWAC-339.*

Holtby, F. (1940), "Rapid Temperature Measurements of Cast Iron with an Immersion Thermocouple," *Trans. Am. Foundrymen's Assoc.* **47**, p. 854.

Holtby, F. (1941), "Rapid Temperature Measurements of Molten Iron and Steel with an Immersion Thermocouple," *Trans. Am. Soc. Metals* **29**, p. 863.

Homewood, C. F. (1941), "Factors Affecting the Life of Platinum Thermocouples," *Temperature, its Measurement and Control in Science and Industry,* vol. 1, Reinhold, New York, p. 1272.

Hoppe, A. W., and P. J. Levine (1966), "Status of High Temperature Thermometry for the Nerva Reactor," *WASH-1067, Paper 32.*

Hornig, D. F., and B. J. O'Keefe (1947), "The Design of Fast Thermopiles and the Ultimate Sensitivity of Thermal Detectors," *Rev. Sci. Instr.* **18**, p. 474.

Hougen, O. A., and B. L. Miller (1923), "How Silica Protection Tubes Cause Contamination of Thermocouples," *Chem. Met. Eng.* **29**, p. 662.

Howard, J. L. (1971), "Error Accumulation in Thermocouple Thermometry," *Fifth Symp. on Temperature, Washington D.C., Paper T-42.*

Hughes, P. C., and N. A. Burley (1962), "Metallurgical Factors Affecting Stability of Nickel-Base Thermocouples," *J. Inst. Metals* **91**, p. 373.

Hukill, W. V. (1941), "Characteristics of Thermocouple Anemometers," *Temperature, its Measurement and Control in Science and Industry,* vol. 1, Reinhold, New York, p. 666.

Hunt, L.B. (1964), "The Early History of the Thermocouple," *Platinum Metals Rev.* **8**, p. 23.

Hunter, M. A., and A. Jones (1941), "Electromotive Force of Alloys in Various Alloy Systems," *Temperature, its Measurement and Control in Science and Industry,* vol. 1, Reinhold, New York, p. 1227.

Huntley, D. J., and C. W. E. Walker (1969), "Magnetic Field Dependence of the Thermoelectric Power at Low Temperatures for Very Dilute Alloys of Fe in Au," *Can. J. Phys.* **47**, p. 805.

Hust, J. G., R. L. Powell, and L. L. Sparks (1971), "Methods for Cryogenic Thermocouple Thermometry," *Fifth Symp. on Temperature, Washington D.C., Paper T-44.*

Hyman, S. C., C. F. Bonilla, and S. W. Ehrlich (1951), "Natural Convection Transfer Processes, I. Heat Transfer to Liquid Metals and Nonmetals at Horizontal Cylinders," *AEC-NYO-564.*

Ihnat, M. E. (1959), "A Jet Engine Thermocouple System for Measuring Temperatures up to 2300°F," *WADC-TR-57-744.*

Ihnat, M. E., and W. C. Hagel (1960), "A Thermocouple System for Measuring Turbine Inlet Temperatures," *Trans. ASME* 82D, p. 81.

Israël, J. (1966), "A Study of the Thermoelectric Properties of Pt-Ru Alloys," *J. Nucl. Mat.* 18, p. 272.

Jaffe, D., and M. C. Hallinan (1961), "Thermocouple System," *U.S. Patent 3,007,988.*

Jamieson, C. P. (1967), "Graphite Thermocouples and Method of Making," *U.S. Patent 3,305,405.*

Jamison, R. E., and T. H. Blewitt (1953), "Behavior of Two Types of Thermocouples Under Pile Irradiation at Low Temperatures," *Rev. Sci. Instr.* 24, p. 474.

Jeffs, A. T. (1967), "The Compatibility of W-26%Re with Thoria and Thoria-Plutonia," *Atomic Energy of Canada, Ltd., AECL-2768.*

Jensen, J. T., J. Klebanoff, and G. A. Haas (1964), "Thermocouple Errors Using Pt/Pt-Rh Thermocouples on Ni Surfaces," *Rev. Sci. Instr.* 35, p. 1717.

Jewell, R. C. (1947), "An Examination of the Microstructure of Contaminated and Embrittled Platinum and Platinum-Rhodium Wires," *J. Iron Steel Inst.* 155, p. 231.

Jewell, R. C., and E. G. Knowles (1951), "Behavior of Platinum/Platinum-Rhodium Thermocouples at High Temperatures," *J. Sci. Instr.* 28, p. 353.

Jewell, R. C., E. G. Knowles, and T. Land (1955), "High Temperature Thermocouple," *Metal Industry* 87, p. 217.

Johnson, E. A., and L. Harris (1933), "Thermoelectric Force of Thin Films," *Phys. Rev.* 44, p. 944.

Johnson, W. P. (1966), "Ultra High Temperature Thermocouple for Aerospace Application," *Proc. 12th Nat. ISA Aerospace Instrumentation Symp., Philadelphia*, p. 275.

Jones, R. V. (1934), "The Design and Construction of Thermoelectric Cells," *J. Sci. Instr.* 11, p. 247.

Jones, T. P. (1968), "The Accuracies of Calibration and Use of I.P.T.S. Thermocouples," *Metrologia* 4, p. 80.

Jones, T. P. (1969), "Reference Tables for Thermocouples," *Natl. Std. Lab. (Australia), Technical Paper No. 26.*

Jones, T. P. (1971), "A Comparison of Fixed Point and Intercomparison Methods of Calibration of Platinum-Platinum, Rhodium Thermocouples in the Temperature Range 0° to 1064°C," *Fifth Symp. on Temperature, Washington, D.C., Paper C-11.*

Kamenetskii, A. B., and N. V. Gul'ko (1965), "Interaction of Thermal Electrodes in Tungsten-Rhenium Thermocouples with Pure Oxides Insulation," *Meas. Tech.* (6), p. 506.

Keesom, W. H., and C. J. Matthijs (1935), "Thermo-emf's at Temperatures from 2.5 to 17.5 K," *Physica* 2, p. 623. Also as *Leiden Comm.* (238b).

Keinath, I. G. (1935), "Nonmetal Thermoelements for Highest Temperatures," *Arch. Tech. Messen* 241, (4), p. T55.

Kellow, M. A., A. N. Bramley, and F. K. Bannister (1969), "The Measurement of Temperatures in Forging Dies," *Int. J. Machine Tool Design and Research* 9, p. 239.

Kelly, M. J. (1960), "Changes in Emf Characteristics of Chromel-Alumel and Platinum/Platinum-Rhodium," *TID-7586,* pt. 1, p. 120.

Kelly, M. J., W. W. Johnston, and C. D. Baumann (1962), "The Effects of Nuclear Radiation on Thermocouples," *Temperature, its Measurement and Control in Science and Industry,* vol. 3, pt. 2, Reinhold, New York, p. 265.

Kendall, D. N., W. P. Dixon, and E. H. Schulte (1967), "Semiconductor Surface Thermocouples for Determining Heat Transfer Rates," *IEEE Trans. Aerospace and Electronic Systems,* **AES-3,** p. 596.

Kent, J. H. (1970), "A Noncatalytic Coating for Platinum-Rhodium Thermocouples," *Combustion and Flame* **14,** p. 279.

King, W. H., Jr., C. T. Camilli, and A. F. Findeis (1968), "Thin Film Thermocouples for Differential Thermal Analysis," *Analytical Chem.* **40,** p. 1330.

Kislyi, P. S., L. S. Terebukh, G. D. Vilyura, L. P. Berchak, and N. A. Skorodinskii (1968), "Stability of the Calibration Curves of Thermocouples in Silicon Carbide Casings," *Meas. Tech.* (9), p. 1210.

Kjaerheim, G. (1965), "Measuring Fuel Rod Center Temperatures," *Euro-Nuclear* **2,** p. 72.

Klapper, J. A., R. Plottel, B. E. Tenzer, and H. Heffan (1965), "High Temperature Corrosion of Constantan Thermocouple Conductors," *Materials Protection* **4,** June, p. 72.

Klein, C. A., and M. P. Lepie (1964), "Operational Performance Characteristics of Pyrolytic Graphite Thermocouples," *Solid State Electronics* **7,** p. 241.

Klein, C. A., M. P. Lepie, W. D. Straub, and S. M. Zalar (1964), "Development of an Ultra-High Temperature Pyrolytic Graphite Thermocouple," *ASD-TDR-63-844.*

Koeppe, W. (1967), "Low Temperature Thermocouple, Gold-Cobalt (2.1 Per Cent at.) versus Chromel," *Cryogenics* **7,** p. 172.

Kohl, W. H. (1967), *Handbook of Materials and Techniques for Vacuum Devices,* Reinhold, New York.

Koike, R., H. Kurokawa, and Y. Iijima (1968), "Thickness Dependence of the Thermoelectromotive Force in Evaporated Bi and Sb Films," *Japan. J. Appl. Phys.* **7,** p. 293.

Kollie, T. G., and R. S. Graves (1966), "The Effect of Cold Working Pt-10Rh/Pt Thermocouples," *WASH-1067, Paper 13.*

Kostkowski, H. J. (1960), "The Accuracy and Precision of Measuring Temperatures Above 1000 K," *Proc. Internat. Symp. on High Temperature Technology,* McGraw-Hill, New York, p. 33.

Kostkowski, H. J., and G. W. Burns (1963), "Thermocouple and Radiation Thermometry Above 900 K," *NASA-SP-31,* p. 13.

Kröckel, O. (1960), "Ein neues Thermoelement für Hochtemperaturmessungen," *Silikattechnik* **11,** p. 108.

Kuether, F. W. (1960), "Uniqueness of Thermal Emf's Measured in Molybdenum, Rhenium, Tungsten, Iron, and Copper," *TID-7586,* pt. 1, p. 23.

Kuether, F. W. (1962), "The Measurement of High Temperature Thermal Emf Characteristics, 1. Emf Instability," *Temperature, its Measurement and Control in Science and Industry,* vol. 3, pt. 2, Reinhold, New York, p. 229.

Kuether, F. W. (1962), "The Measurement of High Temperature Thermal Emf Characteristics, 2, Emf Scatter," *Temperature, its Measurement and Control in Science and Industry,* vol. 3, pt. 2, Reinhold, New York, p. 233.

Kuether, F. W. (1962), "The Thermal Emf Scatter of Rhenium," *Rhenium,* B. W. Gonser, ed., Elsevier, Amsterdam, p. 149.

Kuether, F. W., and J. C. Lachman (1960), "How Reliable are the Two New High Temperature Thermocouples in Vacuum?," *ISA Journal* 7, April, p. 67.

Kuhlman, W. C. (1963, May), "Research and Evaluation of Materials for Thermocouple Application Suitable for Temperature Measurements up to $4500°F$ on the Surface of Glide Re-Entry Vehicles," *ASD-TDR-63-233*.

Kuhlman, W. C. (1963, Sept.), "Status Report of the Investigation of Thermocouple Materials for Use at Temperatures above $4500°F$," *Soc. Automotive Engrs., National Aeronautic and Space Engineering and Manufacturing Meeting, Los Angeles, Paper 750D.*

Kuhlman, W. C. (1965), "Evaluation of Thermal Neutron Induced Errors in the W/W-25Re Thermocouple," *AIAA Propulsion Joint Specialist Conf., Colorado Springs, Colo., Paper 65-563.*

Kuhlman, W. C. (1966), "Thermocouple Research at GE-NMPO, *WASH-1067, Paper 26.*

Kuhlman, W., and W. Baxter (1969), "A 1000 Hour, $2300°C$ Thermocouple Test," *Trans. Am. Nucl. Soc.* 12, p. 319.

Kuhlman, W. C., and W. G. Baxter (1969), "1000 Hour $2300°C$ Thermocouple Stability Tests," *General Electric Co., GEMP-738,* Oct.

Kutzner, K. (1968), "Gold-Iron Thermocouples," *Cryogenics* 8, p. 325.

Lachman, J. C. (1961), "New Developments in Tungsten/Tungsten-Rhenium Thermocouples," *16th Annual ISA Conf. Proc.* 16, pt. 2, Paper 150-LA-61. Also in *ISA Trans.* 1, 1962, p. 340.

Lachman, J. C. (1962), "High Temperature Thermocouples Using Tungsten-Rhenium Alloy Tubing," *Soc. Automotive Engrs., National Aeronautic Meeting, New York, Paper 524D.*

Lachman, J. C., and F. W. Kuether (1960), "Stability of Rhenium/Tungsten Thermocouples in Hydrogen Atmosphere," *TID-7586,* pt. 1, p. 31.

Lachman, J. C., and F. W. Kuether (1960), "Stability of Rhenium-Tungsten Thermocouples in Hydrogen Atmospheres," *ISA Journal* 7, March, p. 67.

Lachman, J. C., and J. A. McGurty (1962), "Thermocouples for $5000°F$ Using Rhenium Alloys," *Rhenium,* B. W. Gonser, ed., Elsevier, Amsterdam, p. 153.

Lachman, J. C., and J. A. McGurty (1962), "The Use of Refractory Metals for Ultra High Temperature Thermocouples," *Temperature, its Measurement and Control in Science and Industry,* vol. 3, pt. 2, Reinhold, New York, p. 177.

Lagedrost, J. F. (1964), "Laboratory Studies Toward Improvement of High Temperature Thermocouples," *TID-7697,* p. 4.4.1.

Lakh, V. I., B. I. Stadnyk, and Yu. B. Kuz'ma (1963), "Thermoelectric Stability of Thermocouples made of Tungsten-Rhenium Alloys at High Temperatures," *High Temp.* 1, p. 267.

Land, T. (1947), "An Investigation of the Embrittlement of Platinum-Rhodium Wire in the Heads of Liquid Steel Pyrometers," *J. Iron Steel Inst.* 155, p. 214.

Lander, J. J. (1948), "Measurements of Thomson Coefficients for Metals at High Temperatures and of Peltier Coefficients for Solid-Liquid Interfaces of Metals," *Phys. Rev.* 74, p. 479.

Lapp, G. B., and D. L. Popova (1964), "Certain Thermometric Properties of Tungsten-Rhenium Thermoelectrodes," *Meas. Tech.* (11), p. 917.

La Rosa, M. (1916), "Thermoelectric Effect of a Carbon-Platinum Couple," *Palermo Nuovo Cimento* 12, p. 284.

Leonard, J. H. (1969), "Conditions Contributing to Radiation Induced Thermocouple Decalibration," *Nucl. Applic.* 6, p. 202.

Leonard, J. H., and D. B. Hunkar (1967), "Direct Radiation Effects on Thermocouples," *Nucl. Applic.* **3**, p. 718.

Lever, R. C. (1959), "Better Thermocouple Alloys Make 1800°C Measurement Feasible," *SAE Journal* **67**, Oct., p. 48.

Levy, G. F., R. R. Fouse, and R. Sherwin (1962), "Operation of Thermocouples under Conditions of High Temperature and Nuclear Radiation," *Temperature, its Measurement and Control in Science and Industry*, vol. 3, pt. 2, Reinhold, New York, p. 277.

Liddiard, F. E., and J. H. Heath (1962), "Temperature Measurement, *Control* 5, Sept., p. 95.

Liermann, J., and S. Tarassenko (1971), "Temperature Measurement on Nuclear Fuel Elements," *Fifth Symp. on Temperature, Washington, D.C., Paper T-15.*

Liquid Steel Temperature Subcommittee (1947), "A Symposium on the Contamination of Platinum Thermocouples, Introduction," *J. Iron Steel Inst.* **155**, p. 213.

Lister, E., and R. P. Harvey (1967), "The Importance of Accurate Testing for Determining Standard High Temperature Properties of Steel," *Conf. on Steels for the Power and Chemical Industry, Czechoslovakia* (paper).

Lohr, J. M. (1920), "Alloys Suitable for Thermocouples and Base Metal Thermoelectric Practice," *Pyrometry*, AIME, New York, p. 181.

Lohr, J. M., C. H. Hopkins, and C. L. Andrews (1941), "The Thermal Electromotive Force of Various Metals and Alloys," *Temperature, its Measurement and Control in Science and Industry*, vol. 1, Reinhold, New York, P. 1232.

Long, E. L., Jr. (1966), "Three Metallographic Observations," *WASH-1067, Paper 3.*

Losana, L. (1940), "Coppie Termoelettriche per Temperature Elevate," *Metallurgia Italiana* **32**, p. 239.

Loscoe, C., and H. Mette (1962), "Limitations in the Use of Thermocouples for Temperature Measurements in Magnetic Fields," *Temperature, its Measurement and Control in Science and Industry*, vol. 3, pt. 2, Reinhold, New York, p. 283.

Lyusternik, V. E. (1963), "Reproducibility of the Calibration of Platinum/Platinum-Rhodium Thermocouples Over a Wide Temperature Range," *High Temp*,1, p. 120.

McCoy, H. E., Jr. (1966), "Influence of $CO-CO_2$ Environments on the Calibration of Chromel-P/Alumel Thermocouples," *WASH-1067, Paper 11.*

MacDonald, D. K. C., W. B. Pearson, and I. M. Templeton (1962), "Thermoelectricity at Low Temperatures, IX. The Transition Metals as Solute and Solvent," *Proc. Roy. Soc. (London)* A**266**, p. 161.

McGeary, R. K. (1952), "Thermocouples of Tantalum, Tungsten, and Molybdenum," *AECD-3541.*

McGurty, J. A., and W. C. Kuhlman (1962), "Tungsten/Tungsten-Rhenium Thermocouple Research and Development," *Soc. Automotive Engrs., National Aeronautic Meeting, New York, Paper 524C.*

MacKenzie, D. J., and M. D. Scadron (1962), "Selection of Thermocouples for High Gas Temperature Measurement," *Soc. Automotive Engrs., National Aeronautic Meeting, New York, Paper 524E.*

McLaren, E. H., and E. G. Murdock (1971), "New Considerations on the Preparation, Properties and Limitations of the Standard Thermocouple for Thermometry," *Fifth Symp. on Temperature, Washington, D.C., Paper T-52.*

McQuilkin, F. R., N. H. Briggs, and E. L. Long, Jr. (1965), "Summary of Experience with High Temperature Thermocouples Used in the ORNL-GCR Program Fuel Irradiation Experiments," *Trans. Am. Nucl. Soc.* **8**, p. 69.

McQuillan, M. K. (Undated), "The Calibration and Use of the Molybdenum-Tungsten Thermocouple at High Temperatures," *Commonwealth of Australia Dept. of Supply and Development, Div. of Aeronautics, SM-134.*

McQuillan, M. K. (1949), "Some Observations on the Behavior of Platinum/Platinum-Rhodium Thermocouples at High Temperatures," *J. Sci. Instr.* **26**, p. 329.

McQuillan, M. K. (1950), Discussion of Troy and Steven (1950) *Trans. Am. Soc. Metals* **42**, p. 1153.

Madsen, P. E. (1951), "The Calibration of Thermocouples Under Irradiation in B.E.P.O.," *United Kingdom Atomic Energy Authority, AERE-M/R-649.*

Malyshev, V. N. (1969), "High Temperature Thermocouple Based on Graphite," *Ind. Lab.* **35**, p. 1064.

Manterfield, D. (1947), "Contamination and Failure of Rare Metal Thermocouples," *J. Iron Steel Inst.* **155**, p. 227.

Marshall, R., L. Atlas, and T. Putner (1966), "The Preparation and Performance of Thin Film Thermocouples," *J. Sci. Instr.* **43**, p. 144.

Martin, C. D., and G. H. Gabbard (1971), "Experience with Type K Thermocouples in an Experimental Nuclear Reactor," *Fifth Symp. on Temperature, Washington D.C., Paper T-22.*

Maykuth, D. J., R. H. Ernst, and H. R. Ogden (1965), "The Thermoelectric Properties of Tantalum Alloys," *Trans. Met. Soc. AIME* **233**, p. 1196.

Medvedeva, L. A., M. P. Orlova, and A. G. Rabinkin (1971), "A Thermocouple for Measuring Low Temperatures," *Cryogenics* **11**, p. 316.

Medvedeva, L. A., M. P. Orlova, I. L. Rogelberg, V. M. Beilin, and N. D. Loutzau (1971), "Investigation on Thermoelectrode Materials for Low Temperature Thermocouples," *Fifth Symp. on Temperature, Washington D.C., Paper T-47.*

Merryman, R. G., and C. P. Kempter (1965), "Precise Temperature Measurement in Debye-Scherrer Specimens at Elevated Temperatures," *J. Am. Ceram. Soc.* **48**, p. 202.

Meservey, R. H. (1967), "The Possible Use of Boron Nitride as a High Temperature Thermocouple Insulator," *Phillips Petroleum Co., IDO-17234*, p. 13.

Metcalf, A. G. (1950), "The Use of Platinum Thermocouples in Vacuo at High Temperatures," *Brit. J. Appl. Phys.* **1**, p. 256.

Miller, J. (1960), "Thermocouple Materials for the Measurement of Temperatures above 1600°C," *Instr. Pract.* **14**, p. 510.

Moeller, C. E. (1962), "Thermocouples for the Measurement of Transient Surface Temperatures," *Temperature, its Measurement and Control in Science and Industry,* vol. 3, pt. 2, Reinhold, New York, p. 617.

Moeller, C. E. (1963), "Special Surface Thermocouples," *Instr. Control Systems* **36**, May, p. 97.

Moeller, C. E., M. Noland, and B. L. Rhodes (1968), "NASA Contributions to Development of Special Purpose Thermocouples, A Survey," *NASA-SP-5050.*

Moffatt, R.J. (1962), "The Gradient Approach to Thermocouple Circuitry," *Temperature, its Measurement and Control in Science and Industry,* vol. 3, pt. 2, Reinhold, New York, p. 33.

Moore, F. D., and R. B. Mesler (1961), "The Measurement of Rapid Surface Temperature Fluctuations during Nucleate Boiling of Water," *Am. Inst. Chem. Engrs. J.* **7**, p. 620.

Moore, J. P., and D. L. McElroy (1966), "Contamination of Platinum-Rhodium Transducers," *WASH-1067, Paper 12.*

Moore, W. C. (1915), "The Thermoelectric Properties of Carbon," *J. Am. Chem. Soc.* **37**, p. 2032.

Morgan, E. S. (1968), "The Effect of Stress on the Thermal Emf of Platinum/Platinum-13 Per Cent Rhodium Thermocouples," *J. Phys. D.* 1, p. 1421.

Morgan, F. H., and W. E. Danforth (1950), "Thermocouples of the Refractory Metals," *J. Appl. Phys.* 21, p. 112.

Morrison, R. D., and R. R. Lachenmayer (1963), "Thin Film Thermocouples for Substrate Temperature Measurement," *Rev. Sci. Instr.* 34, p. 106.

Mortlock, A. J. (1958), "Error in Temperature Measurement Due to the Interdiffusion at the Hot Junction of a Thermocouple," *J. Sci. Instr.* 35, p. 283.

Morugina, S. (1926), "The High Temperature Thermoelectric Force of the W-Ta and W-Mo Thermocouple," *Z. Tech. Phys.* 7, p. 486.

Müller, L. (1930), "Bestimmung des Schmelzpunktes von Chrom mit Thermoelement," *Ann. Physik* 7, p. 48.

Muraoka, H. (1881), "Thermoelectric Powers of Various Forms of Carbon," *Wied. Ann.* 13, p. 307.

Nadler, M. R., and C. P. Kempter (1961), "Thermocouples for Use in Carbon Atmospheres," *Rev. Sci. Instr.* 32, p. 43.

Nielson, I. O. (1970), "Selection of Thermocouples for Temperature Profiling of Semiconductor Diffusion Furnaces," *Solid State Technology* 13, Oct., p. 33.

Nishimura, H. (1958), "Thermocouple and Elements Thereof," *U.S. Patent 2,861,114.*

Nishimura, H. J. (1958), "Investigation of Pt-Mo Alloys," *J. Japan. Inst. Metals* 22, p. 425.

Northover, E. W., and J. A. Hitchcock (1968), "The Effect of Heating on the Local Thermoelectric Power Characteristics of Commercial Thermocouple Wires," *Instr. Pract.* 22, p. 606.

Novak, P. E., and R. R. Asamoto (1965), "Evaluation of Thermocouples for Use to $2600°C$ in Mixed Oxide Fuel," *Trans. Am. Nucl. Soc.* 8, p. 388.

Novak, P. E., and R. R. Asamoto (1966), "An Out-of-Pile Evaluation of W-Re Thermocouple Systems for Use to 4700 Deg. F in PuO_2-UO_2," *General Electric Co., GEAP-5166.*

Novak, P. E., and R. R. Asamoto (1967), "Thermocouple Development for Use in Measuring Oxide Fuel Temperatures In-Pile Above $4000°F$," *Proc. Internat. Symp. on Developments in Irradiation Capsule Technology,* D. R. Hoffman, ed., Joint AEC/GE Meeting.

Novak, P. E., and R. R. Asamoto (1968), "An In-Pile Evaluation of W-Re Thermocouple Systems for Use to $4700°F$ in SEFOR Fuel," *General Electric Co., GEAP-5468.*

Obrowski, W., I. Gessler, and H. J. Hantke (1971), "High Temperature Measurements ($950°C$) in the German Pebble Bed Reactor with NiCr-Ni Thermocouple Assemblies," *Fifth Symp. on Temperature, Washington, D.C., Paper T-16.*

Obrowski, W., and W. Prinz (1962), "Neu bestimmte Grundwerte für die Thermopaarkombination Pt 30% Rh-Pt 6% Rh," *Archiv für das Eisenhüttenwesen* 33, p. 1.

Obrowski, W., and C. Von Seelen (1961), "Thermocouple," *U.S. Patent 2,990,440.*

Oleinikova, L. D., P. P. Oleinikov, and L. I. Trakhtenberg (1969), "Stability of Tungsten-Rhenium Thermocouples VR 5/20 and VR 10/20 in the Range 1450-$2000°C$," *Meas. Tech.* (10), p. 1373.

Olsen, L. O. (1962), "Catalytic Effects of Thermocouple Materials," *Soc. Automotive Engrs., National Aeronautic Meeting, New York, Paper 524G.*

Olsen, L. O. (1963), "Some Recent Developments in Noble Metal Thermocouples," *Soc. Automotive Engrs., National Aeronautic and Space Engineering and Manufacturing Meeting, Los Angeles, Paper 750A.*

Olsen, L. O., and P. D. Freeze (1964), "Reference Tables for the Platinel II Thermocouple," *J. Res. Natl. Bur. Std. (U.S.)* **68C**, p. 263.

Onuma, Y. (1965), "Thermocouple Consisted of Germanium Thin Film Deposited in Vacuum," *Japan. J. Appl. Phys.* **4**, p. 814.

Osann, B., Jr., and E. Schröder (1933), "Temperaturmessungen mit Wolfram-Molybdän Thermoelementen," *Archiv für das Eisenhüttenwesen* **7**, p. 89.

Othmer, D. F., and H. B. Coats (1928), "Measurement of Surface Temperature," *Ind. Eng. Chem.* **20**, p. 124.

Palmer, E. P., and G. H. Turner (1966), "Response of Thermocouple Junctions to Shock Waves," *ATL-TR-66-36.*

Panasyuk, A. D., and G. V. Samsonov (1963), "Thermocouples with Electrodes of Refractory Carbides for Use up to 3000˘C," *High Temp.* **1**, p. 116.

Pearson, W. B. (1954), "Thermocouples Suitable for Use at Low Temperature," *J. Sci. Instr.* **31**, p. 444.

Pesko, R. N., R. L. Ash, S. G. Cupschalk, and E. F. Germain (1971), "Theory and Performance of Plated Thermocouples," *Fifth Symp. on Temperature, Washington D.C., Paper T-40.*

Peters, E. T., and J. J. Ryan (1966), "Comment on the Pressure Dependence of the Emf of Thermocouples," *J. Appl. Phys.* **37**, p. 933.

Pirani, M., and G. von Wangenheim (1925), "A Thermocouple for the Highest Temperatures," *Z. Tech Phys.* **6**, p. 358.

Pletenetskii, G. E. (1967), "Thermocouples for Measuring High Temperatures in a Vacuum," *Instr. and Exper. Tech.* (1), p. 214.

Pletenetskii, G. E., and A. T. Mandrich (1965), "A Study of Thermoelectric Materials in Helium at 1500°C," *Ind. Lab.* **30**, p. 1539.

Pollack, D. D., and D. I. Finch (1962), "The Effect of Cold Working Upon Thermoelements," *Temperature, its Measurement and Control in Science and Industry,* vol. 3, pt. 2, Reinhold, New York, p. 237.

Popper, G. F., and A. E. Knox (1967), "Reactor Fuel Pin Temperatures," *IEEE Trans. Nucl. Sci.* **NS-14**, p. 333.

Popper, G. F., and T. Z. Zeren (1966), "Refractory Oxide Insulated Thermocouple Analysis and Design," *WASH-1067, Paper 21.*

Potter, R. D. (1954), "Calibration of a Nickel-Molybdenum Thermocouple," *J. Appl. Phys.* **25**, p. 1383.

Potter, R. D., and N. J. Grant (1949), "Tungsten-Molybdenum Thermocouples," *Iron Age* **163**, March 31, p. 65.

Potts, J. F., Jr., and D. L. McElroy (1960), "Some Results of Base Metal Thermocouple Research at Oak Ridge National Laboratory," *TID-7586,* pt. 1, p. 1.

Potts, J. F., Jr., and D. L. McElroy (1961), "Thermocouple Research to 1000°C, Final Report, Nov. 1, 1957 through June 30, 1959," *ORNL-2773.*

Potts, J. F., Jr., and D. L. McElroy (1962), "The Effects of Cold Working, Heat Treatment, and Oxidation on the Thermal Emf of Nickel-Base Thermoelements," *Temperature, its Measurement and Control in Science and Industry,* vol. 3, pt. 2, Reinhold, New York, p. 243.

Powell, R. L., M. D. Bunch, and L. P. Caywood (1961), "Low Temperature Thermocouple Thermometry," *Advances in Cryogenic Engineering,* vol. 6, Plenum Press, New York, p. 537.

Powell, R. L., M. D. Bunch, and R. J. Corruccini (1961), "Low Temperature Thermo-couples, 1. Gold-Cobalt or Constantan vs. Copper or 'Normal' Silver," *Cryogenics* **1**, p. 139.

Powell, R. L., and L. P. Caywood, Jr. (1962), "Low Temperature Characteristics of Some Commercial Thermocouples," *Advances in Cryogenic Engineering,* vol. 7, Plenum Press, New York, p. 517.

Powell, R. L., L. P. Caywood, Jr., and M. D. Bunch (1962), "Low Temperature Thermo-couples," *Temperature, its Measurement and Control in Science and Industry,* vol. 3, pt. 2, Reinhold, New York, p. 65.

Powell, R. L., H. M. Roder, and W. M. Rogers (1957), "Low Temperature Thermal Con-ductivity of Some Commercial Coppers," *J. Appl. Phys.* **28**, p. 1282.

Prince, W. R., and W. L. Sibbitt (1965), "High Temperature Reactor Core Thermocouple Experiments," *LA-3336-MS.* Also as *Paper 65-562, AIAA Propulsion Joint Specialist Conf., Colorado Springs, Colo., 1965.*

Pumphrey, W. I. (1947), "The Embrittlement of Chromel and Alumel Thermocouple Wires," *J. Iron Steel Inst.* **157**, p. 513.

Rainey, W. T., Jr., and R. L. Bennett (1963), "Stability of Base Metal Thermocouples in Helium Atmospheres," *ISA Trans.* **2**, p. 34.

Rainey, W. T., Jr., R. L. Bennett, and H. L. Hemphill (1963), "Stability of Chromel-P/ Alumel Thermocouples Sheathed in Stainless Steel," *ORNL-3417,* p. 163.

Rall, D. L. (1960), "Temperature Instrumentation of Metallic Uranium Fuel Pins," *TID-7586,* pt. 1, p. 157.

Ramachandran, S. (1963), "Temperature Measurements in Steel Baths at Elevated Tem-peratures and in Vacuum Systems," *18th Annual ISA Conf., Chicago, Paper 21.3.63.*

Ramachandran, S., and T. R. Acre (1964), "Measuring Molten Steel Temperatures," *ISA Journal* **11**, March, p. 54.

Rauch, W. G. (1954), "Design and Construction of Needle Thermocouples," *Metal Progress* **65**, March, p. 71.

Rechowicz, M., T. Ashworth, and H. Steeple (1969), "The Measurement of Thermal Con-ductivity at Low Temperatures," *Cryogenics* **9**, p. 58.

Reed, R. P., and E. A. Ripperger (1971), "Deposited Thin Film Thermocouples of Micro-scopic Dimensions," *Fifth Symp. on Temperature, Washington D.C., Paper L-15.*

Reeve, L., and A. Howard (1947), "Fracture of Platinum and Platinum-13% Rhodium Wires used in the Immersion Thermocouple," *J. Iron Steel Inst.* **155**, p. 216.

Reichardt, F. A. (1963), "Measurement of High Temperatures Under Irradiation Con-ditions," *Platinum Metals Rev.* **7**, p. 122.

Reid, R. J., and W. M. Gross (1957), "Fast Response Thermocouple," *Electronic Design* **5**, July, p. 98.

Rhys, D. W. (1969), "Precious Metal Thermocouples," *Metals and Materials* **3**, March, p. 95.

Rhys, D. W., and P. Taimsalu (1969), "Effect of Alloying Additions on the Thermoelectric Properties of Platinum," *Engelhard Industries Tech. Bull.* **10**, p. 41.

Richards, D. B., L. R. Edwards, and S. Legvold (1969), "Thermocouples in Magnetic Fields," *J. Appl. Phys.* **40**, p. 3836.

Ridgway, R. R. (1939), "Thermocouple," *U.S. Patent 2,152,153.*

Robertson, D. (1956), "Neutron Sensitive Thermopile for Reactor Applications, Progress Report No. 1," *AECU-3416.*

Roeser, W. F. (1941), "Thermoelectric Thermometry," *Temperature, its Measurement and Control in Science and Industry,* vol. 1, Reinhold, New York, p. 180.

Roeser, W. F., and A. I. Dahl (1938), "Reference Tables for Iron-Constantan and Copper-Constantan Thermocouples," *J. Res. Natl. Bur. Std. (U.S.)* **20**, p. 337.

Roeser, W. F., A. I. Dahl, and G. J. Gowens (1935), "Standard Tables for Chromel-Alumel Thermocouples," *J. Res. Natl. Bur. Std. (U.S.)* **14**, p. 239.

Roeser, W. F., and S. T. Lonberger (1958), "Methods of Testing Thermocouples and Thermocouple Materials," *Natl. Bur. Std. Circular No. 590.*

Roeser, W. F., and H. T. Wensel (1933), "Reference Tables for Platinum to Platinum-Rhodium Thermocouples," *J. Res. Natl. Bur. Std. (U.S.)* **10**, p. 275.

Roeser, W. F., and H. T. Wensel (1941), "Methods of Testing Thermocouples and Thermocouple Materials," *Temperature, its Measurement and Control in Science and Industry,* vol. 1, Reinhold, New York, p. 284.

Roeser, W. F., and H. T. Wensel (1941), Appendix to *Temperature, its Measurement and Control in Science and Industry,* vol. 1, Reinhold, New York, p. 1293.

Roess, L. C., and E. N. Dacus (1945), "The Design and Construction of Rapid Response Thermocouples for Use as Radiation Detectors in Infrared Spectrographs," *Rev. Sci. Instr.* **16**, p. 164.

Rohn, W. (1924), "Thermoelektrische Untersuchungen an Nickellegierungen," *Z. Metallkunde* **16**, p. 297.

Rohn, W. (1927), "Metalle und Legierungen für Thermoelemente zur Messung hoher Temperaturen," *Z. Metallkunde* **19**, p. 138.

Rolin, M., and G. Poupa (1967), "Pyrométrie de contact aux Températures supérieures à $1500°C$," *Proc. 7th Internat. Conf., States of Matter under the Extreme Effects of Very High and Very Low Temperatures and Pressures, Paris,* p. 153.

Rosebury, F. (1965), *Handbook of Electron Tube and Vacuum Techniques,* Addison-Wesley, Reading, Mass.

Rosenbaum, R. L. (1968), "Some Properties of Gold-Iron Thermocouple Wire," *Rev. Sci. Instr.* **39**, p. 890.

Rosenbaum, R. L. (1969), "Some Low Temperature Thermometry Observations," *Rev. Sci. Instr.* **40**, p. 577.

Rosenbaum, R. L. (1970), "A Survey of Some Secondary Thermometers for Possible Applications at Very Low Temperatures," *Rev. Sci. Instr.* **41**, p. 37.

Rosenbaum, R. L., R. R. Oder, and R. B. Goldner (1964), "Low Temperature Thermoelectric Power of Gold-Iron vs. Copper Thermocouples," *Cryogenics* **4**, p. 333.

Ross, C. W. (1961), "Effect of Thermal Neutron Irradiation on Thermocouples and Resistance Thermometers," *IRE Trans. Nucl. Sci.* NS-8, p. 110.

Rudnitskii, A. A. (1959), "Thermoelectric Properties of the Noble Metals and Their Alloys," *AEC-tr-3724.*

Rudnitskii, A. A., and I. I. Tyurin (1956), "The Investigation and Choice of Alloys for High Temperature Thermocouples," *Russian J. Inorg. Chem.* **1**, p. 1074. Also as *AERE Lib./Trans.* 798, 1958.

Rudnitskii, A. A., and I. I. Tyurin (1960), "New Alloys for High Temperature Thermocouples," *Russian J. Inorg. Chem.* **5**, p. 192.

Sagoschen, J. (1961), "Temperaturmessung mit Platinmetall-Thermoelementen," *Metall.* **15**, p. 34.

Salzano, F. J. (1966), "Stability of a Rhenium-Graphite Thermocouple," *WASH-1067, Paper 34.*

Samsonov, G. V., and P. S. Kislyi (1967), *High Temperature Nonmetallic Thermocouples and Sheaths,* Consultants Bureau, the Plenum Publishing Corp., New York.

Samsonov, G. V., P. S. Kislyi, and A. D. Panasyuk (1962), "Thermoelectric Properties of Thermocouples with High Melting Solid Electrodes," *Meas. Tech.* (10), p. 810.

Sanders, V. D. (1958), "Review of High Temperature Immersion Thermal Sensing Devices for In-Flight Engine Control," *Rev. Sci. Instr.* **29**, p. 917.

Sandfort, R. M., and E. J. Charlson (1968), "Calibration Curves for Gold-Nickel and Silver-Nickel Thin Film Thermocouples," *Solid State Electronics* **11**, p. 635.

Sasagawa, K., H. Hotta, and T. Omuro (1949), "Research on Measurement of the Temperature of Molten Steel," *Japan Sci. Rev.* **1**, p. 179.

Sata, T., and R. Kiyoura (1963), "Studies on the Tungsten Furnace and its Temperature Measurement by W/W-Re Thermocouple up to 2700°C," *Bull. Tokyo Inst. Tech.* (51), p. 39.

Schneider, J. F. (1967), "Noble Metal Thermocouple Having Base Metal Compensating Leads," *U.S. Patent 3,328,209.*

Schriempf, J. T., and A. I. Schindler (1965), "Comments on 'Low Temperature Thermoelectric Power of Gold-Iron versus Copper Thermocouples'," *Cryogenics* **5**, p. 174.

Schriempf, J. T., and A. I. Schindler (1966), "Low Temperature Thermoelectric Power of Gold-Iron versus Silver Normal Thermocouples," *Cryogenics* **6**, p. 301.

Schulte, E. H., and R. F. Kohl (1970), " A Transducer for Measuring High Heat Transfer Rates," *Rev. Sci. Instr.* **41**, p. 1732.

Schulze, A. (1933), "Thermoelements at High Temperatures," *Z. Ver. Deut. Ing.* **77**, p. 1241.

Schulze, A. (1939), "Metallic Materials for Thermocouples," *J. Inst. of Fuel* **12**, p. S41. Also, in *Chem. Zeit.* **62**, 1938, pp. 308, 485; and, with additional data, as "Über metallische Werkstoffe für Thermoelemente,"*Metall.* **18**, 1939, pp. 249, 271, 315.

Schwarz, E., *British Patent Specifications 578187, 578188.*

Schwiegerling, P. E. (1971), "The Effects of Processing Variables on the Thermoelectric Properties of Metal Sheathed Iron-Constantan Thermocouples," *Fifth. Symp. on Temperature, Washington D.C., Paper T-33.*

Scott, R. B. (1941), "The Calibration of Thermocouples at Low Temperatures," *Temperature, its Measurement and Control in Science and Industry,* vol. 1, Reinhold, New York, p. 206. Also in *J. Res. Natl. Bur. Std. (U.S.)* **25**, 1940 p. 459.

Selman, G. L. (1971), "On the Stability of Metal Sheathed Noble Metal Thermocouples," *Fifth Symp. on Temperature, Washington D.C., Paper T-27.*

Selman, G. L., and R. Rushforth (1971), "The Stability of Metal Sheathed Platinum Thermocouples," *Platinum Metals Rev.* **15**, p. 82.

Sessions, C. E. (1966), "Metal Reactions with Al_2O_3," *WASH-1067, Paper 10.*

Shaffer, P. T. B. (1963), Discussion of R. J. Bertodo (1963), *Proc. Inst. Mech. Engrs.* **177**, p. 615.

Sharp, J. D. (1963), "Continuous Temperature Measurement of Molten Steel," *Platinum Metals Rev.* **7**, p. 90. Also in *J. Metals* **15**, 1963, p. 902.

Shenker, H., J. I. Lauritzen, Jr., R. J. Corruccini, and S. T. Lonberger (1955), "Reference Tables for Thermocouples," *Natl. Bur. Std. Circular 561.*

Shepard, R. L., H. S. Pattin, and R. D. Westbrook (1956), "A High Temperature Boron-Graphite/Graphite Thermocouple," *Bull. Am. Phys. Soc.* **1**, p. 119.

Shoens, C. J., and J. W. Shortall (1953), "Thermoelectric Calibration of Zirconium-Constantan and Zirconium-Alumel Thermocouples," *Livermore Research Lab., LRL-62*.

Sibley, F. S. (1960), "Effect of Environment on the Stability of Chromel-Alumel Thermocouples," *TID-7586*, pt. 1, p. 16.

Sibley, F. S., N. F. Spooner, and B. F. Hall, Jr. (1968), "Aging in Type K Couples," *Instr. Tech.* **15**, July, p. 53.

Simons, J. P., C. G. Hamstead, and E. J. Burton (1953), "The Tungsten-Molybdenum Thermocouple for Immersion Pyrometry," *J. Iron Steel Inst.* **175**, p. 402.

Sims, C. T., C. M. Craighead, R. I. Jaffee, D. N. Gideon, W. W. Kleinschmidt, W. E. Nexsen, Jr., G. B. Gaines, F. C. Todd, C. S. Peet, D. M. Rosenbaum, R. J. Runck, and I. E. Campbell (1956), "Investigations of Rhenium," *WADC-TR-54-371, Supplement 1*.

Sims, C. T., G. B. Gaines, and R. I. Jaffee (1959), "Refractory Metal Thermocouples Containing Rhenium," *Rev. Sci. Instr.* **30**, p. 112.

Sims, C. T., and R. I. Jaffee (1956), "Further Studies of the Properties of Rhenium Metal," *AIME Trans.* **206**, p. 913.

Sirota, A. M., and B. K. Mal'tsev (1959), "A Gold-Platinum Thermocouple," *Meas. Tech.* (8), p. 611.

Skaggs, S. R., W. A. Ranken, and A. J. Patrick (1967), "Decalibration of a Tungsten-Tungsten 25% Rhenium Thermocouple in a Neutron Flux," *LA-3662*.

Slack, G. A. (1961), "Thermal Conductivity of CaF_2, MnF_2, CoF_2, and ZnF_2 Crystals," *Phys. Rev.* **122**, p. 1451.

Slaughter, J. I., and J. L. Margrave (1962), "Temperature Measurement," *SSD-TDR-62-140*.

Slaughter, J. I., and J. L. Margrave (1967), "Temperature Measurement," *High Temperature Materials and Technology*, I. E. Campbell and E. M. Sherwood, eds., Wiley, New York, p. 717.

Smith, R. A., F. E. Jones, and R. P. Chasmar (1957), *The Detection and Measurement of Infrared Radiation*, Oxford University Press, Oxford.

Sosman, R. B. (1910), "The Platinum-Rhodium Thermoelement from 0 to 1755°," *Am. J. Sci.* **30**, (175), p. 1.

Sparks, L. L., and W. J. Hall (1969), "Cryogenic Thermocouple Tables, Part III, Miscellaneous and Comparison Material Combinations," *Natl. Bur. Std. (U.S.)* 9721.

Sparks, L. L., and R. L. Powell (1967), "Cryogenic Thermocouple Thermometry," *Natl. Bur. Std. (U.S.)* R427.

Sparks, L. L., and R. L. Powell (1971), "Reference Data for Thermocouple Materials Below the Ice Point," *Fifth Symp. on Temperature, Washington D.C., Paper T-45*.

Sparks, L. L., and R. L. Powell (1971), "Laboratory Method for Assessing Homogeneity and Interchangeability of Thermocouple Wires," *Materials Research and Stds.* **11**, Aug., p. 19.

Sparks, L. L., R. L. Powell, and W. J. Hall (1968), "Cryogenic Thermocouple Tables," *Natl. Bur. Std. (U.S.)* 9712.

Sparks, L. L., R. L. Powell, and W. J. Hall (1970), "Cryogenic Thermocouple Research," *ISA Trans.* **9**, p. 243.

Spooner, N. F., and J. M. Thomas (1955), "Longer Life for Chromel-Alumel Thermocouples," *Metal Progress* **68**, Nov., p. 81.

Stadnyk, B. I., and G. V. Samsonov (1964), "Thermocouples for Measuring High Temperatures," *High Temp.* **2**, p. 573.

Stafsudd, O., and N. B. Stevens (1968), "Thermopile Performance in the Far Infrared," *Appl. Optics* **7**, p. 2320.

Stanforth, C. M. (1962), "Problems Encountered Using Ir-Rh/Ir Thermocouples for Measuring Combustion Gas Temperatures," *Soc. Automotive Engrs., National Aeronautic Meeting, New York, Paper 524F.*

Starr, C. D., and T. P. Wang (1969, 1971), "Thermocouples and Extension Wires," *Instr. Control Systems* **42**, Oct., 1969, p. 111. *Fifth Symp. on Temperature, Washington D.C., 1971, Paper T-1.*

Stauss, H. E. (1941), "Platinum and Pyrometry," *Temperature, its Measurement and Control in Science and Industry,* vol. 1, Reinhold, New York, p. 1267.

Stern, J., S. Edelman, P. Freeze, and D. Thomas (1970), "Preliminary Results of Noble Metal Thermocouple Research Program, 1000-2000°C," *NASA-CR-72646.*

Steven, G. (1956), "High Temperature Thermocouples," *High Temperature Technology,* I. E. Campbell, ed., Wiley, New York, p. 357.

Stimson, H. F. (1961), "International Practical Temperature Scale of 1948, Text Revision of 1960," *J. Res. Natl. Bur. Std. (U.S.)* **65A**, p. 139. Also as *Natl. Bur. Std. Monograph 37,* 1961.

St. Pierre, D. S. (1960), "An Internally Wound Platinum-Rhodium Furnace," *Am. Ceram. Soc. Bull* **39**, p. 264.

Stromberg, H. D., and D. R. Stephens (1966), "Effect of Pressure on the Temperature Calibration of Mo/Mo-50% Re Thermocouples to 1000°C and 60 Kilobars," *Univ. Calif., Lawrence Radiation Laboratory, UCRL-14993.*

Strong, J. (1938), *Procedures in Experimental Physics,* Prentice-Hall, Englewood Cliffs, N.J.

Strongin, M., J. M. Dickey, H. H. Farrell, T. F. Arns, and G. Hrabak (1971), "A Cryostat for Low Energy Electron Diffraction Work at Liquid Helium Temperatures," *Rev. Sci. Instr.* **42**, p. 311.

Stroud, R. C. (1971), "Compatibility of Thermocouple Alloys and Components in Various Atmospheres," *Fifth Symp. on Temperature, Washington D.C., Paper T-30.*

Studennikov, Yu. A., and G. E. Erkovich (1965), "Compensation Wires for VR 5/20 Thermocouples," *Ind. Lab.* **31**, p. 469.

Sturm, W. J., and R. J. Jones (1954), "Applications of Thermocouples to Target Temperature Measurement in the Internal Beam of a Cyclotron," *Rev. Sci. Instr.* **25**, p. 392.

Svec, H. J. (1952), "Behavior of Platinum/Platinum-Rhodium Thermocouples at High Temperatures," *J. Sci. Instr.* **29**, p. 100.

Szaniszlo, A. J. (1969), "Thermal Electromotive Force Change for 87Pt 13Rh/Pt Thermocouples in 1530 K, 10^{-8} Torr Environment for 3700 Hours," *NASA-TN-D-5287.*

Tanaka, M., and K. Okada (1937), "Three Element Thermocouple for Precision Temperature Measurements," *Electrotech. Lab., Tokyo, Research No. 404.*

Terebukh, L. S., G. D. Vilyura, L. P. Berchak, L. T. Struk, P. S. Kislyi, and M.A. Kuzenkova (1968), "Tungsten-Rhenium Thermocouples for Periodically Checking the Temperature in Furnaces for the Fire Refining of Copper," *High Temp.* **6**, p. 1045.

Thielke, N. R., and R. L. Shepard (1960), "High Temperature Thermocouples Based on Carbon and its Modifications," *TID-7586,* pt. 1, p. 44.

Thomas, D. B. (1963), "Studies on the Tungsten-Rhenium Thermocouple to 2000°C," *J. Res. Natl. Bur. Std. (U.S.)* **67C**, p. 337.

Thomas, D. B., and P. D. Freeze (1971), "The Effects of Catalysis in Measuring the Temperature of Incompletely Burned Gases with Noble Metal Thermocouples," *Fifth Symp. on Temperature, Washington D.C., Paper T-38.*

Thornburg, D. D., and C. M. Wayman (1969), "Thermoelectric Power of Vacuum Evaporated Au-Ni Thin Film Thermocouples," *J. Appl. Phys.* **40**, p. 3007.

Toenshoff, D. A., and E. D. Zysk (1971), "Material Preparation and Fabrication Techniques for the Production of High Reliability Thermocouple Devices," *Fifth Symp. on Temperature, Washington D.C., Paper T-12.*

Trauger, D. B. (1960), "LITR Experiments Relating to Rhenium-Tungsten and Platinum/Platinum-Rhodium Thermocouples," *TID-7586,* pt. 1, p. 130.

Trogolo, J. A. (1971), "Ceramic Thermocouple Insulation," *Fifth Symp. on Temperature, Washington D.C., Paper T-10.*

Troy, W. C., and G. Steven (1950), "The Tungsten-Iridium Thermocouple for Very High Temperatures," *Trans. Am. Soc. Metals,* **42**, p. 1131.

Tseng, Y., A. Robertson, and E. D. Zysk (1968), "Platinum-Molybdenum Thermocouple," *Engelhard Industries Tech. Bull.* **9**, p. 77.

Tseng, Y., S. Schnatz, and E. D. Zysk (1970), "Tungsten 3 Rhenium vs Tungsten 25 Rhenium Thermocouple, Some Recent Developments," *Engelhard Industries Tech. Bull.* **11**, p. 12.

Tyushev, V. S., and O. V. Shelud'ko (1969), "Germanium Film Thermocouple," *Ind. Lab.* **35**, p. 1378.

Ubbelohde, A. R., L. C. F. Blackman, and P. H. Dundas (1959), "A Graphite-Graphite Thermocouple for High Temperatures," *Chem. and Ind.* (19), p. 595.

Van Liempt, J. A. M. (1929), "Thermal Elements for High Temperature in Reducing Atmosphere," *Recueil des Travaux Chimiques* **48**, p. 585.

Vedernikov, M. V. (1969), "The Thermoelectric Powers of Transition Metals at High Temperatures," *Advances in Physics* **18**, p. 337.

Vigor, C. W., and J. R. Hornaday, Jr. (1962), "A Thermocouple for Measurement of Temperature Transients in Forging Dies," *Temperature, its Measurement and Control in Science and Industry,* vol. 3, pt. 2, Reinhold, New York, p. 625.

Villamayor, M. (1968), "Study of Thermocouples for Measurement of High Temperatures," *NASA-TT-F-11623.* Also as *CEA-R-3182,* 1967.

Vines, R. F., and E. M. Wise (1941), *The Platinum Metals and their Alloys,* The International Nickel Co., New York.

Vonkeman, G. H. (1970), *The Thermocouple $Pt/Cr_2O_3/Pt$ and its Use in Determining Melting Curves at Higher Temperatures,* Ph.D. Diss., Univ. Utrecht, Netherlands. Available from *Nat. Tech. Info. Service, N70-43034.*

Von Middendorff, A. (1971), "Thermocouples at Low Temperatures in High Magnetic Fields," *Cryogenics* **11**, p. 318.

Wagner, H. J., and J. C. Stewart (1962), "An Exploration of the Thermoelectric Properties of Certain Cobalt Base Alloys," *Temperature, its Measurement and Control in Science and Industry,* vol. 3, pt. 1, Reinhold, New York, p. 245.

Walker, B. E., C. T. Ewing, and R. R. Miller (1962), "Thermoelectric Instability of Some Noble Metal Thermocouples at High Temperatures," *Rev. Sci. Instr.* **33**, p. 1029; **34**, p. 1456.

Walker, B. E., C. T. Ewing, and R. R. Miller (1965), "Study of the Instability of Noble Metal Thermocouples in Vacuum," *Rev. Sci. Instr.* **36**, p. 601.

Walker, B. E., C. T. Ewing, and R. R. Miller (1965), "Instability of Refractory Metal Thermocouples," *Rev. Sci. Instr.* **36**, p. 816.

Walters, P. A. (1966), "Thermopile Improvement," *NASA-CR-80768.*

Wasan, V. P., and C. L. Gupta (1967), "Thermocouples for High Temperature Measurement," *Engelhard Industries Tech. Bull.* **8**, p. 77.

Watson, G. G. (1966), "Techniques for Measuring Surface Temperature, 2.," *Instr. Pract.* **20**, p. 335.

Watson, H. L., and H. Abrams (1928), "Thermoelectric Measurement of Temperatures above 1500°C," *Trans. Am. Electrochem. Soc.* **54**, p. 19.

Weber, R. L. (1950), *Heat and Temperature Measurement,* Prentice-Hall, New York, p. 64.

Weiss, J., and J. Königsberger (1909), "The Thermoelectric Power of the Copper-Carbon Couple," *Phys. Zeit.* **10**, p. 956.

Welch, J. H. (1956), "The System CaO-MgO-Al$_2$O$_3$-SiO$_2$, Phase Equilibrium Relationships along the Joint Anorthite-Spinel," *J. Iron Steel Inst.* **183**, p. 275.

Wenzl, M., and F. Morawe (1927), From Stahl und Eisen, May 1927, p. 867, as reviewed in *Supplement to the Engineer* **3**, Aug. 1927, p. 119.

West, E. D. (1960), "Techniques in Calorimetry, I. A Noble Metal Thermocouple for Differential Use," *Rev. Sci. Instr.* **31**, p. 896.

Westbrook, R. D., and R. L. Shepard (1960), "Carbon/Carbon-Boron Thermocouple," *U.S. Patent 2,946,835.*

White, R. M. (1968), "The Determination of the Elevated Temperature Seebeck Coefficients of Common Thermocouple Materials Referenced to Platinum," *Sandia Corp., SC-RR-68-538.*

Wichers, E. (1962), "The History of Pt 27," *Temperature, its Measurement and Control in Science and Industry,* vol. 3, pt. 1, Reinhold, New York, p. 259.

Wilhelm, H. A., H. J. Svec, A. I. Snow, and A. H. Danne (1948), "High Temperature Thermocouples," *AECD-3275.*

Williams, R. K., M. Veeraburus, and W. O. Philbrook (1968), "Radial Heat Flow Thermal Conductivity Apparatus for Measurements on Sulfide and Telluride Melts," *Rev. Sci. Instr.* **39**, p. 1104.

Wintle, C. A., and K. J. Salt (1967), "Improved Accuracy of Temperature Measurement with Thermocouples," *United Kingdom Atomic Energy Authority, AEEW-R-522.*

Wisely, H. R. (1955), "Thermocouples for Measurement of High Temperatures," *Ceramic Age* **66**, July, p. 15.

Woldman, N. E. (1949), *Materials Engineering of Metal Products,* Reinhold, New York.

Wong, P., and P. S. Schaffer (1961), "Development of a Tungsten-Molybdenum Immersion Thermocouple for Measuring Molten Metal Temperatures in Vacuum," *WAL-TR-831.13/1.*

Wood, V. E. (1967), "Thermal Neutron Transmutation Effects on W/W-26Re Thermocouples," *J. Appl. Phys.* **38**, p. 1756.

Worsham, H. J., Jr., and H. D. Warren (1969), "An Investigation into the Effects of In-Pile Radiation and Temperature on the Leadwires of Chromel-Alumel Thermocouples," *IEEE Nucl. Sci. Trans.* **NS-16**, p. 188.

Yust, C. S., and L. L Hall (1966), "The Pt$_{60}$Rh$_{40}$/Pt$_{80}$Rh$_{20}$ Thermocouple," *WASH-1067, Paper 15.*

Zuikov, N. V., A. A. Tsvetaev, and M. E. Bardyukov (1965), "Thermocouple to Measure Temperatures up to 2500 K," *High Temp.* **3**, p. 756.

Zysk, E. D. (1960), "Developments on High Temperature Thermocouples using Noble Metals," *Engelhard Industries Tech. Bull.* 1, p. 8.

Zysk, E. D. (1962), "Platinum Metal Thermocouples," *Temperature, its Measurement and Control in Science and Industry,* vol. 3, pt. 2, Reinhold, New York, p. 135.

Zysk, E. D. (1963), "A Review of Recent Work with the Platinel Thermocouple," *Engelhard Industries Tech. Bull.* 4, p. 5.

Zysk, E. D. (1964), "Noble Metals in Thermometry, Recent Developments," *Engelhard Industries Tech. Bull.* 5, p. 69.

Zysk, E. D. (1967), "Tungsten-Rhenium Alloy Thermocouple and Matching Lead Wires," *U.S. Patent 3,296,035.*

Zysk, E. D. (1968), "Noble Metal Thermocouple having Base Metal Leads," *U.S. Patent 3,372,062.*

Zysk, E. D., and E. E. Osovitz (1970), "Thermocouple with Tungsten-Rhenium Alloy Leg Wires," *U.S. Patent 3,502,510.*

Zysk, E. D., E. E. Osovitz, and R. W. Stephans (1969), "Noble Metal Thermocouple having Base Metal Compensating Leads," *U.S. Patent 3,451,859.*

Zysk, E. D., and A. R. Robertson (1971), "Newer Thermocouple Materials," *Fifth Symp. on Temperature, Washington D.C., Paper T-3.*

Zysk, E. D., and D. A. Toenshoff (1967), "Calibration of Refractory Metal Thermocouples," *Engelhard Industries Tech. Bull.* 7, p. 137.

Zysk, E. D., D. A. Toenshoff, and J. Penton (1963), "Tungsten 3 Rhenium vs Tungsten 25 Rhenium, a New High Temperature Thermocouple," *Engelhard Industries Tech. Bull.* 3, p. 130.

Index